ADVANCES IN MULTI-PHOTON PROCESSES AND SPECTROSCOPY

ADVANCES IN MULTI-PHOTON PROCESSES AND SPECTROSCOPY

Volume 21

Edited by

S. H. Lin
National Chiao-Tung University, TAIWAN
Institute of Atomic and Molecular Sciences, TAIWAN
and Arizona State University, USA

A. A. Villaeys
Institute de Physique et Chimie des
Matériaux de Strasbourg, FRANCE

Y. Fujimura
Tohoku University, JAPAN

World Scientific

NEW JERSEY · LONDON · SINGAPORE · BEIJING · SHANGHAI · HONG KONG · TAIPEI · CHENNAI

Published by

World Scientific Publishing Co. Pte. Ltd.

5 Toh Tuck Link, Singapore 596224

USA office: 27 Warren Street, Suite 401-402, Hackensack, NJ 07601

UK office: 57 Shelton Street, Covent Garden, London WC2H 9HE

British Library Cataloguing-in-Publication Data
A catalogue record for this book is available from the British Library.

Library of Congress Control Number: 86643116

Advances in Multi-Photon Processes and Spectroscopy — Vol. 21
ADVANCES IN MULTI-PHOTON PROCESSES AND SPECTROSCOPY
(Volume 21)

ISBN 978-981-4518-33-8

Typeset by Stallion Press
Email: enquiries@stallionpress.com

PREFACE

In view of the rapid growth in both experimental and theoretical studies of multi-photon processes and multi-photon spectroscopy of molecules, it is desirable to publish an advanced series that contains review articles readable not only by active researchers, but also by those who are not yet experts and intend to enter the field. The present series attempts to serve this purpose. Each chapter is written in a self-contained manner by experts in the area so that readers can grasp the content without too much preparation.

This volume consists of six chapters. The first chapter presents the results of both theoretical and experimental studies of "Vibrational and Electronic Wavepackets Driven by Strong Field Multi-photon Ionization". First, basic theoretical ideas essential to understanding multiphoton ionization and laser control of molecules are described. Secondly, experimental techniques for molecular control such as phase-dependent dissociation, photon locking and spatial hole burning are explained by taking halogenated methanes (CH_2BrI, CH_2I_2) as a model system.

The second chapter deals with the results of experimental studies on "Orientation-selective Molecular Tunneling by Phase-controlled Laser Fields". After the basic properties of tunneling ionization (TI) of atoms and molecules are introduced, the experimental results of directionally asymmetric TI of CO, OCS, iodohexane, and bromochloroethane, which are induced by $\omega + 2\omega$ laser pulses, are presented.

The third chapter presents experimental and theoretical results of "Reaction and Ionization of Polyatomic Molecules Induced by Intense Laser Pulses". The emphasis is on ionization rates, resonance effects, dissociative ionization and Coulomb explosion of polyatomic molecules such as cyclopentanone (C_5H_8O), which are induced by intense *fs*-laser fields.

The fourth chapter presents the reviews of experimental studies on "Ultrafast Internal Conversion of Pyrazine via Conical Intersection". Pyrazine is one of the typical azabenzenes undergoing ultrafast S_2–S_1 internal conversion through conical intersection. In this chapter, experimental results of femtosecond internal conversion of pyrazine, that are observed in real time using a time-resolved photoelectron imaging method with a time resolution of 22 fs are presented. The method enables us to obtain a time–energy map of the photoelectron angular anisotropy as well.

The fifth chapter deals with the theoretical studies of "Quantum Dynamics in Dissipative Molecular Systems". Dissipation is essential in condensed phase systems. Femtosecond time-resolved spectroscopy applied to photosynthetic antenna in proteins manifests as quantum beats, which indicates the quantum nature of the system. The timescale of the protein environment memory is found to be comparable to that of the energy transfer. For such a system, traditional perturbative Markovian quantum dissipation theories are inadequate. The reviews of theoretical studies in the nonperturbative and non Markovian treatments are presented on the basis of the hierarchical equation of motion approach.

The sixth chapter presents the results of the theoretical and computational studies of "First-principle Calculations for Laser Induced Electron Dynamics in Solids". Electron dynamics in a crystalline solid induced by strong ultrashort laser pulses is totally different from that observed in atoms and molecules. The basic principles and restrictions for treating electrons in crystalline solids are described. Time-dependent Kohn-Sham equation in a unit cell is solved based on the time-dependent density functional theory. The present theory and computational method provide the most comprehensive description for the interactions of strong and ultrashort laser pulses with solids.

The editors wish to thank all the authors for their important contributions to Advances in Multi-photon Processes and Multiphoton Spectroscopy Vol. 21. It is hoped that the collection of topics in this volume will be useful not only to active researchers but also to other scientists and graduate students in scientific research fields such as chemistry, physics, and material science.

CONTENTS

3.   Reaction and Ionization of Polyatomic Molecules
Induced by Intense Laser Pulses   105

D. Ding, C. Wang, D. Zhang, Q. Wang, D. Wu and S. Luo

*K. Yabana, Y. Shinohara, T. Otobe, Jun-Ichi Iwata
and George F. Bertsch*

CHAPTER 1

VIBRATIONAL AND ELECTRONIC WAVEPACKETS DRIVEN BY STRONG FIELD MULTIPHOTON IONIZATION

P. Marquetand[*,¶], T. Weinacht[†], T. Rozgonyi[‡],
J. González-Vázquez[§], D. Geißler[†] and L. González[*]

We present basic theoretical ideas underlying multiphoton ionization and laser control of molecules. Approaches to describe molecular electronic structure, spin-orbit coupling, dynamic Stark shifts, dressed states, and multiphoton excitations are shortly reviewed. Control techniques such as phase-dependent dissociation, photon locking, and spatial hole burning are explained and illustrated exemplarily using halogenated methanes (CH_2BrI, CH_2I_2) as model systems. Theoretical approaches are compared with experiments and the complex signals resulting from phenomena like electronic wavepackets are elucidated and understood. Hence, we show how strong-field control concepts developed for simple systems can be transferred to more complex ones and advance our ability to control molecular dynamics.

1.1. Introduction

The development of intense ultrafast lasers over the past two decades has led to dramatic advances in our ability to follow molecular dynamics on femtosecond and attosecond timescales.[1-5] Furthermore, intense ultrafast lasers not only provide the means to study electronic and nuclear dynamics,

[*]Institute of Theoretical Chemistry, University of Vienna, Währinger Straße 17, 1090 Vienna, Austria
[†]Department of Physics, Stony Brook University, Stony Brook, New York 11794, USA
[‡]Institute of Materials and Environmental Chemistry, Research Centre for Natural Sciences, Hungarian Academy of Sciences, Pusztaszeri út 59-67, Budapest, HU-1025, Hungary
[§]Instituto de Química Física Rocasolano, CSIC, C/Serrano 119, 28006 Madrid, Spain
[¶]Email: philipp.marquetand@univie.ac.at

but also allow for influencing their evolution. While several control schemes have been described and implemented in diatomic molecules,[6–16] this chapter focuses on following and controlling vibrational dynamics in a family of small polyatomic molecules — the halogenated methanes CH_2XY (X, Y = I, Br, Cl...). Being small enough to allow for high-level *ab initio* electron-structure calculations, but offering sufficient complexity for chemical relevance (e.g. atmospheric chemistry, bond selective dissociation, conical intersections), and presenting a homologous series for laser selective chemistry, these molecules are ideal for testing different control schemes, characterizing electronic wavepackets generated via strong-field ionization (SFI), and for implementing strong field control over bond breaking.

In this chapter, we outline many of the basic physical and computational principles underlying the dynamics and control, and discuss several measurements and calculations which illustrate them. The first few sections deal with solving the time-independent and time-dependent Schrödinger equation (TISE and TDSE, respectively) for polyatomic molecular systems via *ab initio* electronic structure theory and wavepacket propagations. The following sections give a brief and simple discussion of basic principles required to understand strong field control, including AC Stark shifts, multiphoton transitions, dressed states and SFI. After these basic ideas are introduced, we discuss the ideas and implementation of photon locking, spatial hole burning, and phase-dependent dissociation. The term "photon locking" (or "optical paralysis"),[17–23] is used to describe the mixing (or dressing) of two potential energy surfaces in order to lock a vibrational wavepacket in position. "Hole burning" (also termed *r*-dependent excitation or "Lochfrass")[24,25] uses strong field excitation to reshape a vibrational wavepacket by population transfer in a spatially narrow window. Similar approaches, using position dependent ionization or strong field driven AC Stark shifts, have been used to create or reshape molecular wavepackets in diatomic molecules.[26–31] Other works using strong fields focused on using light-dressed states to control the branching ratio in dissociation.[32–36] Finally, we show how pump-probe spectroscopy of vibrational dynamics in conjunction with electronic structure and quantum dynamics can be used to characterize electronic wavepackets generated via strong-field molecular ionization. We conclude with a discussion of future perspectives.

1.2. Theoretical Concepts

1.2.1. *The time-independent Schrödinger equation and its implications on dynamics*

Understanding a chemical reaction induced and/or controlled by strong laser pulses at the molecular level, requires the simulation of motion of nuclei under the influence of an external electric field. This can be done either classically or quantum-mechanically. In both cases the forces governing the motion of the nuclei must be determined either *a priori* or on-the-fly for all the relevant configurations. A fundamental approximation here is the Born–Oppenheimer (BO) approximation which — based on the huge difference between masses of electrons and nuclei — assumes that the motion of nuclei and that of the electrons are separable, i.e., electrons adjust to a nuclear configuration abruptly and the nuclei move in an effective field of the electrons, expressed by the electronic ground- or excited-state potential, $V(\underline{R})$. ($\underline{R}$ represents the coordinates of the nuclei and accordingly, we will denote a vector as $\underline{a}$ and a matrix as $\underline{\underline{A}}$ in the following.) Apart from the most simple cases (when one can use some analytic functions for $V(\underline{R})$ fitted to spectroscopic data), the forces acting on the nuclei — being usually simply the gradient of $V(\underline{R})$ — are obtained by solving the time-independent Schrödinger equation (TISE) for the electronic system. Treating the motions of both the electrons and the nuclei quantum-mechanically, the system is described by the total wavefunction, $\Psi(\underline{r}, \underline{R})$, written as

$$\Psi(\underline{r}, \underline{R}, t) = \sum_n \psi_n(\underline{R}, t)\phi_n(\underline{r}, \underline{R}), \qquad (1.1)$$

where $\phi_n(\underline{r}, \underline{R})$ are the eigenfunctions of the electronic TISE,

$$\hat{H}_e\phi_n(\underline{r}, \underline{R}) = V_n(\underline{R})\phi_n(\underline{r}, \underline{R}) \qquad (1.2)$$

and $\psi_n(\underline{R})$ are the nuclear wavefunctions in electronic states n. In Eq. (1.2), $\hat{H}_e$ is the Hamilton operator of the whole system for fixed nuclei. In the semiclassical dipole approximation the motion of the nuclei in the presence of an external electric field, $\varepsilon(t)$, is governed by the time-dependent Schrödinger equation (TDSE). In matrix form,

$$i\hbar\frac{\partial}{\partial t}\underline{\psi} = (\underline{\underline{T}} + \underline{\underline{V}} - \underline{\underline{\mu}}\varepsilon(t))\underline{\psi}, \qquad (1.3)$$

where $\underline{\underline{T}}$ is the kinetic energy operator for the nuclei and $\underline{\underline{\mu}}$ is the dipole matrix with elements $\underline{\mu}_{nm}$ defined as

$$\underline{\mu}_{nm} = \langle \phi_n | e\underline{r} | \phi_m \rangle, \qquad (1.4)$$

with e being the electron charge and the elements of the ψ vector are the $\psi_n(\underline{R})$ wavefunctions. In the BO approximation $\underline{\underline{V}}$ is a diagonal matrix with elements being the $V_n(\underline{R})$ solutions of Eq. (1.2). In the following, we assume the laser field polarization and the dipole moment vector to be aligned and hence, neglect their vectorial properties.

Depending on the size of the system and the required accuracy, solving the electronic TISE can be very time-consuming so that this is the bottleneck from the point of view of the simulation time. Since quantum-dynamical simulations require the solution of the electronic TISE for several nuclear configurations and also the solution of the nuclear TDSE can become very costly, such computations can only be performed in reduced dimensionality. Therefore, the first step is to choose coordinates appropriate to the process under investigation (e.g., bond length in case of a dissociation). Using normal-mode coordinates (e.g., in case of a bending motion) can simplify the numerical treatment of the nuclear TDSE considerably. Normal-mode coordinates are determined by diagonalizing the mass-weighted Hessian matrix, the elements of which are the second derivatives of the potential energy, V, with respect to Cartesian displacement coordinates of the nuclei from their equilibrium configuration. Having determined the $V(\underline{R})$ on a grid in the space of the selected coordinates, the eigenfunctions belonging to $V(\underline{R})$ can be determined by solving the TISE for the nuclei, e.g., by the Fourier-Grid-Hamiltonian method.[37] In most cases, the lowest-energy vibrational eigenfunction represents the initial nuclear wavefunction for the quantum-dynamical simulations.

In the following, we consider two different approaches to solve the electronic TISE: (i) the wavefunction-based (*ab initio*) methods and (ii) the density-based (Density Functional Theory, DFT) methods. *Ab initio* methods start from the Hartree–Fock (HF) wavefunction, which is an anti-symmetrized product (a Slater-determinant) of one-electron spin-orbitals (molecular orbitals, MO).[38] These orbitals are products of a spatial part and the spin-eigenfunction. In practice, the spatial orbitals are constructed by linear combinations of atomic orbitals (LCAO), the so-called basis set. At the HF level of theory, the electronic Hamiltonian is a sum of one-particle operators, the so-called Fock-operators. In this theory, the expansion coefficients in the LCAO are determined by solving the TISE in a self-consistent iterative procedure (called self-consistent field (SCF)) which — according to the variational principle — results in the lowest-energy electronic eigenfunction. Such a wavefunction fulfills the Pauli exclusion principle for fermions, it accounts for the correlation between electrons of the same spin. However, methods based on one Slater determinant, as HF, cannot describe the correlated motion of electrons completely and are generally not appropriate to describe excited electronic states. The correlation effects missing from HF-theory can be classified as static and dynamic correlations. The former arises e.g., in bond dissociations or when different electronic excited states get close in energy. Description of such situations requires multiconfigurational wavefunctions, which are linear combinations of different Slater determinants, obtained by promoting one or more electrons from occupied MOs of the reference Slater determinant to unoccupied ones. Typical multiconfigurational wavefunctions include only the most important determinants. In the complete active space self-consistent field (CASSCF) method,[39] these configurations are constructed by all possible arrangements of electrons within a properly selected small set of orbitals, the so-called active orbitals, and the coefficients of these configurations (CI coefficients) are optimized together with the MO coefficients in the SCF procedure. In this framework, excited electronic states are computed in the state-averaged CASSCF (SA-CASSCF) procedure, where the average energy of a prescribed number of electronic states is minimized in the SCF.

While multiconfigurational procedures like SA-CASSCF account for static or long range electron correlation effects, they are usually not good enough to obtain spectroscopic accuracy, since they do not include enough dynamic correlation. This type of correlation is the result of the instantaneous repulsion of electrons, i.e., the fact that they avoid each other during their motion. The multi-reference configuration interaction (MRCI) method[40] offers a solution to this problem. It relies on a multiconfigurational wavefunction (typically a CASSCF wavefunction) as a reference function and includes further single, double, etc. CI excitations on top of it. This highly accurate method suffers however from two shortcomings: First, it is applicable only to relatively small molecules due to its huge computational cost and second, it is not size-consistent and therefore requires further corrections, such as Davidson correction.[41]

A popular alternative to MRCI is the CASPT2 method in which second-order perturbation theory is applied to a SA-CASSCF reference wavefunction.[42, 43] This method, whose success still strongly depends on the adequate choice of the active space, is also able to provide good estimates of electronic energies, while it is — due to its considerably lower computational costs — applicable to larger systems than MRCI. Computing different electronic states separately by CASPT2 can however result in nonorthogonal electronic wavefunctions. This is an unphysical solution of the nondegenerate eigenvalue problem and can cause inaccurate results when the electronic states are close in energy and their wavefunctions are mixed with one another at the SA-CASSCF level. A solution to this problem is offered by the multistate version of the CASPT2 method[44] in which an effective Hamiltonian is constructed from the single-state solutions and diagonalized producing new wavefunctions and accurate excitation energies.

In addition to methods based on multiconfigurational wavefunctions, there are several other approaches based on a single reference description of the ground state, which are used to compute excited electronic states. Such methods are e.g., the configuration interaction singles (CIS)[45] which is simple and fast but often cannot even provide qualitatively correct results[46] or the equation of motion coupled cluster (EOM-CC) methods[47] which can produce accurate excitation energies but only at a high excitation level and therefore for an extraordinary computational cost.

Among methods based on the single-reference ground-state description, the most popular for computing excited states is an extension of DFT: the time-dependent density functional theory (TDDFT). The original DFT is based on the finding, that all molecular electronic properties (including energy and wavefunction) are uniquely determined by the electronic ground-state electron density.[48, 49] The energy of the electronic ground state is a functional of the ground-state electron density and the true density minimizes this energy functional. The form of this functional is, however, unknown. Plenty of high quality functionals have been developed, the difference among them being the way they construct the so-called exchange-correlation part of the functional. One of the most widely used functional is the B3LYP.[50–52]

Solving the frequency-dependent polarizability equations,[53,54] TDDFT is able to determine the excitation energies and transition dipole moments (TDFs) without explicitly determining the electronic states. In contrast to CASSCF-based methods, TDDFT is much more simple to use as it does not require the — sometimes tedious — construction of a proper active space. In addition, it is much faster and applicable to much larger systems than multiconfigurational methods. The main disadvantage of the method is that the single configuration for the ground state does not allow a correct description of double or higher excitations and the method is unable to treat degenerate situations correctly. Furthermore, in contrast to multiconfigurational methods, where the accuracy of the computations can be systematically improved by increasing the number of configurations (e.g., increasing the active space in CASSCF), in case of DFT there is no universal functional equally good for any system and there is no way to systematically improve the accuracy.

1.2.2. *Spin-orbit coupling and diabatic vs. adiabatic states*

The electron spin, which cannot be classically understood, is an intrinsic angular momentum of the electron. It gives rise to a magnetic moment, which can interact with the magnetic field that is created when the electron orbits the nucleus. This interaction is consequently termed spin-orbit coupling (SOC). The spin arises naturally from a relativistic description of the electron, as in Dirac's theory.[55] However, the Dirac equation is a single-particle equation and a many-body equation has not yet been derived.[56] Thus, approximate Hamiltonians are used for the electron system, e.g., the Dirac–Coulomb–Breit (index DCB) operator:

$$\hat{H}_{\text{DCB}} = \sum_{i=1}^{n} \hat{h}_D(i) + \sum_{i=1}^{n} \sum_{i<j}^{n} \left[\frac{1}{r_{ij}} - \hat{B}_{ij} \right], \qquad (1.5)$$

which — besides the well-known Coulomb interaction $\frac{1}{r_{ij}}$ — contains the Dirac single-particle Hamiltonian $\hat{h}_D$ and the Breit operator $\hat{B}$.[57] The latter accounts mainly for SOC while the former accommodates predominantly other, scalar relativistic effects.

The Breit operator can be transformed to the so-called Breit–Pauli operator,[58,59] which in principle can be solved numerically but contains many two-electron integrals. The latter can be approximated in the spin-orbit mean-field operator approach, where a single particle is treated in a mean-field of all the others (similar to HF theory).[60] The numerical implementation is called atomic mean field integrals (AMFI).[61]

The Dirac Hamiltonian $\hat{h}_D$ contains the so-called Dirac matrices, which are of size 4×4, and consequently, the corresponding wavefunction has to be a 4-component vector (called a 4-spinor).[57] They contain contributions of electronic as well as positronic type. Note that the latter are not related to positrons but rather negative energies and thus, unphysical artifacts.[62] According to Douglas–Kroll theory, these electronic and positronic states can be decoupled by a unitary transformation of the Dirac Hamiltonian, where the latter is then in a block-diagonal form.[63] The method was later adapted for numerical implementation by Hess.[64] The Douglas–Kroll–Hess method is nowadays used in many quantum chemistry packages and provides scalar relativistic corrections at low computational cost.[62]

The relativistic corrections change the potential shape while the SOCs introduce off-diagonal elements in the Hamiltonian matrix. In cases when the potentials get close in energy, these off-diagonal elements have the effect that population is transferred between the different electronic states. The same effect can also be introduced by other nonadiabatic couplings, e.g., the commonly evaluated kinetic couplings (also called derivative couplings), which can be transformed to potential couplings. In all these cases, the BO approximation breaks down and a single potential is not enough to describe the dynamics of the system.[21]

As indicated above, the off-diagonal elements can be in the potential part as well as in the kinetic part of the Hamiltonian. Different representations exist, where these couplings are transformed in order to ease their application in different methods. First, we focus in a representation that makes use of adiabatic potentials. In this so-called adiabatic picture, the potential matrix is diagonal and the eigenvectors of this matrix are the adiabatic eigenfunctions of the system, which form the basis for the expansion of the total wavefunction. Note, that the term "adiabatic" is sometimes used in a sloppy way to describe simply the output of electronic structure calculations based on the BO approximation. Typical programs yield adiabatic potentials as long as SOC is not considered. As soon as SOCs are computed, they are usually given as potential couplings and the term "adiabatic" for the corresponding potentials (i.e., the diagonal elements of the nondiagonal matrix) is not appropriate anymore. Only after a diagonalization of the potential matrix, the adiabatic picture is obtained. Also the laser interaction can be regarded as a potential coupling. If the matrix including these dipole couplings is diagonalized, the result are the so-called field-dressed states (see Sec. 1.2.6).

If the potential matrix is not diagonal, we speak of a diabatic representation. However, there is no unique definition of a diabatic picture and great care has to be taken in order to avoid misunderstandings. Here, we shortly concentrate on a special case. Each diabatic representation has its respective basis functions. If the wavefunction character of every eigenfunction is retained and thus, the basis functions are time-independent, we speak of a spectroscopic representation since spectroscopic properties very much relate to the wavefunction character. Sometimes, this special case is also termed as "the" diabatic representation. Note that all representations can in principle be interconverted by similarity transformations (although often difficult in practice).

1.2.3. *Nuclear time-dependent Schrödinger equation*

In this section, we describe the possible ways to solve the TDSE, see Eq. (1.3), focusing on vibrational one-dimensional systems (i.e., we use R instead of $\underline{R}$), where the kinetic operator can be described as a diagonal matrix within the BO approximation with elements:[65]

$$\hat{T} = -\frac{\hbar^2}{2}\frac{\partial}{\partial R}g\frac{\partial}{\partial R} = \frac{\hat{p}g\hat{p}}{2}, \qquad (1.6)$$

where g represents the inverse of the moved mass in the coordinate R. This coordinate can be an internal coordinate (e.g., a bond distance) or a collective coordinate (e.g., a normal mode). Depending on the definition of R, g can be a function (e.g., the bending angle), or a constant (e.g., the reduced mass belonging to some normal vibrational mode of a polyatomic molecule),[65] in which case the kinetic operator is just $\hat{T} = g\hat{p}^2/2$. Applying this definition of the kinetic operator to the TDSE, we obtain a series of equations coupled by the electric field,

$$i\hbar\frac{\partial}{\partial t}\psi_m(R,t) = (V_m + \frac{\hat{p}g\hat{p}}{2})\psi_m(R,t) - \sum_n \mu_{nm}\varepsilon(t)\psi_m(R,t). \quad (1.7)$$

This set of equations can be solved by applying the Hamiltonian to the wavefunction, where the procedure mainly depends on the chosen basis. On the one hand, it is possible to choose the i vibrational eigenfunctions of the field-free Hamiltonian for the m electronic potentials, V_m, as a basis,

$$\psi_m(R,t) = \sum_i c_{i,m}(t)\phi_{i,m}(R), \quad (1.8)$$

where $(V_m + \hat{T})\phi_{i,m}(R) = E_{i,m}\phi_{i,m}(R)$ and $c_{i,m}(t)$ are the amplitudes at every time. By inserting Eq. (1.8) in Eq. (1.7) and projecting on $\phi_{j,n}$, we obtain

$$i\hbar\frac{\partial}{\partial t}c_{j,n}(t) = \sum_i [E_{i,m}^{j,n} - \mu_{i,m}^{j,n}\varepsilon(t)]c_{i,m}(t), \quad (1.9)$$

where the integrals of the time-independent Hamiltonian elements are $E_{i,m}^{j,n} = \int \phi_{j,n}^*(\hat{V} + \hat{T})\phi_{i,m}dR$, which in case of orthogonal eigenfunctions are just $E_{i,m}^{j,n} = E_{i,m}\delta_{ij}\delta_{mn}$. Similarly, $\mu_{i,m}^{j,n}$ are the matrix elements of the dipole moment, which can be related to the electronic dipole moment μ_{nm} as $\mu_{i,m}^{j,n} = \int \phi_{j,n}^*\mu_{nm}\phi_{i,m}dR$. If there is no electric field, this equation can be analytically solved and the time evolution of the coefficients is just

$$c_{i,m}(t) = c_{i,m}(0)\exp\left(-\frac{i}{\hbar}E_{i,m}t\right). \quad (1.10)$$

However, neither the calculation of the eigenfunctions nor the solution including an electric field are analytic beyond the harmonic model and the numerical approximation requires many basis functions to solve the above equation.[21]

On the other hand, we can work directly in the one-dimensional grid R, where V_m and μ_{mn} are directly defined. In this case, the problem is the definition of the kinetic operator, which is readily applied in the momentum space but not in the coordinate space. However, the nuclear wavefunction can be easily transformed to the momentum space by a Fourier transform.

In the momentum representation, the kinetic operator is diagonal,

$$\psi_m(R, t) = \sum_i c_{i,m}(t) |r_i\rangle \xrightarrow{FT} \psi_m(p, t) = \sum_i c'_{i,m}(t) |p_i\rangle, \quad (1.11)$$

where $|r_i\rangle$ and $|p_i\rangle$ are basis sets with zeros in all the grid points and 1 when $R = r_i$ and $p = p_i$, respectively. Using this definition, the application of the kinetic operator to the wavefunction is simple. For example,

$$-\frac{1}{2}g\hat{p}^2 \sum_i c'_{i,m}(t) |p_i\rangle = -\sum_i c'_{i,m}(t)\frac{g}{2}p_i^2 |p_i\rangle \quad (1.12)$$

and similarly in the potential part, where $V_m |r_i\rangle = V_m(r_i) |r_i\rangle$ and $\mu_{mn} |r_i\rangle = \mu_{mn}(r_i) |r_i\rangle$.

1.2.3.1. *Second-order differentiator*

One of the simplest methods to solve the TDSE on a grid is the second-order differentiator (SOD) method.[66] In this approach, the wavefunction at time $t + \Delta t$ is expanded in a second-order Taylor expansion:

$$\psi(t + \Delta t) = \psi(t) + \Delta t \frac{\partial}{\partial t}\psi(t) + \frac{\Delta t^2}{2}\frac{\partial^2}{\partial t^2}\psi(t), \quad (1.13)$$

where we can obtain the temporal derivative of ψ using Eq. (1.7). To avoid the application of the Hamiltonian twice, it is possible to modify the propagator, so that[21]

$$\psi(t + \Delta t) = \psi(t - \Delta t) + 2\Delta t \frac{\partial}{\partial t}\psi(t). \quad (1.14)$$

The main problem of this propagator is the numerical instability. Since the propagator is not unitary, the time-step Δt should be very small to assure the conservation of the norm.

1.2.3.2. *Split-operator method*

A more elaborated propagator is the so-called split-operator (SO) technique.[67–69] In this method, we integrate the TDSE from Eq. (1.10) and arrive at a solution as:

$$\psi(t + \Delta t) = \exp\left(-\frac{i}{\hbar}\hat{H}\Delta t\right)\psi(t) = \exp\left(-\frac{i}{\hbar}(\hat{W} + \hat{T})\Delta t\right)\psi(t),$$

(1.15)

where $\hat{W} = \hat{V} - \mu E$ represents the potential part of the Hamiltonian, including the field interaction, which is represented in coordinate space, and $\hat{T}$ is the kinetic part that should be applied in momentum space. As they are not represented on the same grid, it is not possible to apply the exponential including both potential and kinetic parts at the same time and they have to be split into two terms. The problem is that $\hat{W}$ and $\hat{T}$ do not commute, and it is not exact to describe the exponential term as the multiplication of two noncommuting ones, i.e., $\exp(\hat{H}) \neq \exp(\hat{W})\exp(\hat{T})$. This problem is solved in the SO by splitting one of the parts in two. For example, splitting the potential part,

$$\psi(t + \Delta t) \approx \exp\left(-\frac{i}{\hbar}\hat{W}\frac{\Delta t}{2}\right)\exp\left(-\frac{i}{\hbar}\hat{T}\Delta t\right)\exp\left(-\frac{i}{\hbar}\hat{W}\frac{\Delta t}{2}\right)\psi(t).$$

(1.16)

The application of the potential part is very simple even if it is not diagonal, e.g., when the electric field couples two electronic states. In that case, the operators are 2×2 matrices, where every matrix element depends on R represented by the grid points r_i. The potential part of the propagation can be easily carried out after a diagonalization of the $\underline{\underline{W}}$ matrix at every r_i,

$$\exp\left(-\frac{i}{\hbar}\underline{\underline{W}}\frac{\Delta t}{2}\right)\underline{\psi} = \underline{\underline{Z}}\exp\left(-\frac{i}{\hbar}\underline{\underline{Z}}^{\dagger}\underline{\underline{W}}\,\underline{\underline{Z}}\frac{\Delta t}{2}\right)\underline{\underline{Z}}^{\dagger}\underline{\psi}$$

$$= \underline{\underline{Z}}\exp\left(-\frac{i}{\hbar}\underline{\underline{D}}\frac{\Delta t}{2}\right)\underline{\underline{Z}}^{\dagger}\underline{\psi},$$

(1.17)

where $\underline{\underline{D}}$ is a diagonal matrix containing the eigenvalues of $\underline{\underline{W}}$ and $\underline{\underline{Z}}$ is the unitary transformation matrix containing the corresponding eigenvectors. In contrast to the SOD, the SO is unitary and very stable. However, the kinetic operator in the exponent cannot be applied using Fourier transform when g depends on the coordinate.

1.2.4. *Stark shifts*

Electric fields are able to shift electronic potentials, which is known as the Stark effect. It is observed as a shift of the molecule's spectral lines, when the molecule is put in a constant electric (direct current DC) field.[70] This phenomenon can also be witnessed when applying an (alternating current) AC field, e.g. a laser field. It is then termed dynamic Stark effect or Autler–Townes effect.[71] The theoretical background is best explained with a simple model. Imagine a two-level system (see Fig. 1), consisting of a ground state $|g\rangle$ and an excited state $|e\rangle$. The energy difference between these two states is $\omega_{eg} = \omega_e - \omega_g = -\omega_{ge}$ (where we use atomic units, i.e., $\hbar = 1$) and the laser frequency is ω_0, which may be detuned by Δ. Therefore, we have $|\Delta| = |\omega_{eg} - \omega_0|$. The laser field is defined as $\varepsilon_{\text{env}}(t)\cos(\omega_0 t)$, where $\varepsilon_{\text{env}}(t)$ is the envelope function of the laser pulse.

We use the interaction picture (index I), i.e., the ground state potential is shifted up by the energy of one photon (dashed horizontal line in Fig. 1). We define energy zero halfway between the shifted ground-state and the excited-state potential energy. Thus, we arrive at the following Hamiltonian[21]

$$\hat{\mathbf{H}}_I(t) = \hbar \begin{pmatrix} -\dfrac{\Delta}{2} & -\dfrac{\mu_{eg}\varepsilon_{\text{env}}(t)}{2\hbar} \\ -\dfrac{\mu_{eg}\varepsilon_{\text{env}}(t)}{2\hbar} & \dfrac{\Delta}{2} \end{pmatrix}. \tag{1.18}$$

Here, μ_{eg} is the transition dipole moment (TDM) between states g and e. We define

$$\chi(t) = \frac{\mu_{eg}}{2\hbar}\varepsilon_{\text{env}}(t), \tag{1.19}$$

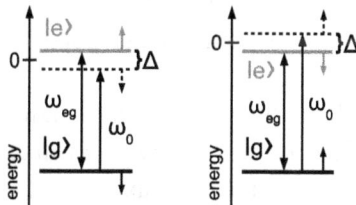

Fig. 1. Stark effect in a two-level system: Negative detuning (left panel) and positive detuning (right panel) lead to different shifts of the involved levels as indicated by the small arrows. See text for more details.

which is just the coupling between the states given in units of a frequency. The definition of the generalized Rabi frequency

$$\Omega(t) = \sqrt{\Delta^2 + \chi^2(t)}, \tag{1.20}$$

is easily rationalized by diagonalizing the above Hamiltonian (Eq. (1.18)). We obtain the eigenvalues

$$E_\pm(t) = \pm\hbar\frac{\Omega(t)}{2}. \tag{1.21}$$

These eigenvalues represent the field-dressed potentials, i.e., the Stark shift of the potentials in time.

It follows from the formalism that the interaction-picture states ($|e\rangle$ and the dashed line in Fig. 1) always "repel" each other. Consequently, $|g\rangle$ is shifted downwards in energy and $|e\rangle$ upwards if the laser frequency is smaller than the energy gap between the states ($\omega_0 < \omega_{eg}$). If the laser frequency is larger than the energy gap ($\omega_0 > \omega_{eg}$), $|g\rangle$ is shifted upwards and $|e\rangle$ downwards, as indicated by the small arrows in Fig. 1.

We can look at two limiting cases: (1) A small detuning is regarded, where $\chi \gg \Delta$. Then, we find that

$$\Omega(t) \approx \chi, \tag{1.22}$$

i.e., the potentials are shifted proportional to the field envelope. This change of the potentials means that the two states mix and consequently, population transfer between the two states takes place, at least temporarily. (2) We consider a large detuning, where $\chi \ll \Delta$ and consequently:

$$\Omega(t) \approx \Delta, \tag{1.23}$$

i.e., the field does not dress the potential substantially. Hence, the two states do not couple considerably and no significant population transfer takes place between them.

To get some idea about the potential shifts induced by different fields, we describe a simple Gedanken experiment: Assume a system with a TDM of 1 a.u. (atomic unit), a field with peak intensity of 1 TW/cm^{-2} (field strength: 0.0053 a.u.) and a detuning of 0.2 eV (0.0073 a.u.). The maximum shift is then 0.01 eV and thus, very small. If the intensity of the field is increased to 100 TW/cm^{-2} (field strength: 0.0533 a.u.), then the potentials experience a large shift of 0.55 eV, respectively. Such a strong field usually leads to ionization in real molecules. These numbers nevertheless indicate the effects to be expected in molecular systems.

1.2.5. *Multi- vs. single-photon transitions*

If the laser field is strong enough (in other words, if the number of photons in a unit volume is high enough), there is a finite probability that more than one photon is absorbed simultaneously by the same molecule. In this case, net population can be transferred between states separated energetically close to some multiple n of the photon energy, $\hbar\omega_0$, by the simultaneous absorption of the n photons. These multiphoton transitions are mediated by other, off-resonant states.

In order to describe such processes, we have to solve the TDSE for the electrons. Similarly to the case of the nuclear TDSE in Eqs. (1.8) and (1.10), we can write the time evolution of the electronic wavefunction as a linear combination of the eigenstates,

$$|\Psi(t)\rangle = \sum_{k=g,m,e} a_k(t)e^{-i\omega_k t}|k\rangle. \tag{1.24}$$

Here, $a_k(t)$ are the state amplitudes for the states $|k\rangle$, where the index k runs over g, m, and e. In this case, g and e stand for the initial (ground) and final (excited) electronic states, respectively, separated roughly by $n\hbar\omega_0$ energy, while m refers to the other states, that we will call intermediate states. The state amplitudes a_k and frequencies ω_k are related to the nuclear wavefunctions ψ_k and electronic state energies V_k as $a_k(t) = \psi_k(t)e^{-i\omega_k t}$ and $\hbar\omega_k = V_k$, respectively.

The energy of the intermediate states need not necessarily be between those of states g and e, but the further an intermediate state is from being in resonance with l photons ($l < n$) to the ground state, the less it can mediate the multiphoton transition from g to e.

In order to capture the basic features of the multiphoton transitions, we start with the simplest multiphoton transition, a two-photon absorption (TPA), which is sketched in Fig. 2, and we introduce some approximations. First, those states, k, that are far off-resonant with respect to a single-photon transition from state g (i.e., $|\omega_{kg} - \omega_0| \gg 0$) can be ignored, since

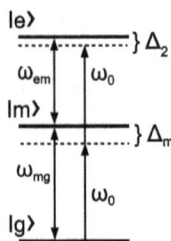

Fig. 2. TPA in a system with ground state g, excited state e and intermediate state m. The excited state is detuned by Δ_2 from the two-photon resonance and the intermediate state by Δ_m from the one-photon resonance of the laser with frequency ω_0.

there is no effective transfer of population to these states — as it is clear from the previous section — and one can keep only those states, m, in the Hamiltonian that are dipole-coupled to both e and g states and are the closest to the single-photon resonance with state g. However, we assume that even these intermediate states are well detuned, i.e., they fulfill the following inequality for the envelope $\varepsilon_{\text{env}}(t)$ of the field:

$$\frac{\partial}{\partial t}\varepsilon_{\text{env}}(t) \ll |\omega_{\{mg,em\}} - \omega_0|. \tag{1.25}$$

This condition is similar to the one usually applied in the slowly-varying-envelope approximation and it means that the intermediate level m is out of the bandwidth range of the photon energy (see Fig. 3). Under these conditions, the differential equations for the intermediate states can be integrated by parts and the result for the state amplitude $a_m(t)$ can be substituted into the TDSE of the state amplitudes $a_g(t)$ and $a_e(t)$.

Defining the two-photon detuning as $\Delta_2 = \omega_{eg} - 2\omega_0$ and using an interaction picture, similar to Eq. (1.18), the TDSE can be reduced to the following approximate simple form, provided that the two-photon detuning is small:

$$i\begin{pmatrix} \dot{a}_g(t) \\ \dot{a}_e(t) \end{pmatrix} = \begin{pmatrix} \omega_g^{(s)}(t) & \chi_2^*(t)e^{-i\Delta_2 t} \\ \chi_2(t)e^{i\Delta_2 t} & \omega_e^{(s)}(t) \end{pmatrix} \begin{pmatrix} a_g(t) \\ a_e(t) \end{pmatrix}, \tag{1.26}$$

where

$$\chi_2(t) = -\sum_m \frac{\mu_{em}\mu_{mg}}{(2\hbar)^2} \frac{(\varepsilon_{\text{env}}(t))^2}{\Delta_m} = \tilde{\chi}_2(\varepsilon_{\text{env}}(t))^2. \tag{1.27}$$

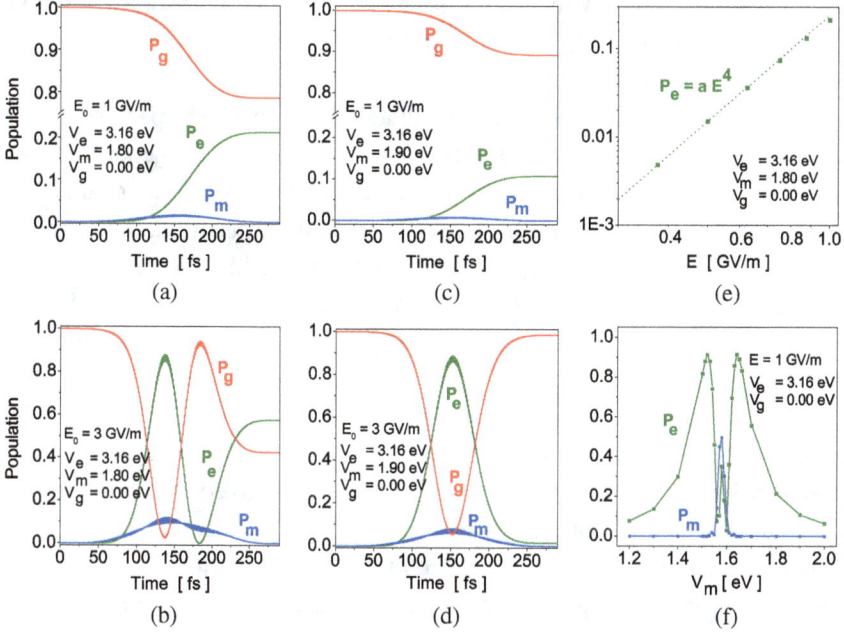

Fig. 3.   TPA with a single intermediate state. The photon energy and the pulse duration are always set to 1.58 eV and 100 fs, respectively. The pulse is centered at 150 fs. Panels (a) and (b) respectively show population dynamics for weak and strong laser fields in case of small detuning, while panels (c) and (d) display similar results in case of large detuning. Panels (e) and (f) show final state populations vs. peak field strength and detuning of the intermediate state, respectively.

Here, $\Delta_m = \omega_{mg} - \omega_0$ is the detuning of the intermediate state m, and χ_2 is the two-photon Rabi frequency. The details of the above procedure, called adiabatic elimination of intermediate states, are given in Ref. 72. As a result of this procedure one ends up with an equation of motion, Eq. (1.26), that has the similar form to that of a single photon absorption in a two-level system. In contrast to the single-photon absorption, however, in the present case the diagonal terms, $\omega_{\{g,e\}}^{(s)}(t)$ are field- and thus time-dependent, i.e., we have Stark shifts even if the two-photon detuning, Δ_2, is zero:

$$\omega_{\{e,g\}}^{(s)}(t) = -\sum_m \frac{\mu_{\{e,g\}m}^2}{2\hbar^2} |\varepsilon_{\text{env}}(t)|^2 \frac{\omega_{m\{e,g\}}}{\omega_{m\{e,g\}}^2 - \omega_0^2} = \tilde{\omega}_{\{e,g\}}^{(s)} |\varepsilon_{\text{env}}(t)|^2.$$

$$(1.28)$$

A further difference with respect to the single-photon absorption is that the two-photon Rabi frequency is proportional to the square of the field amplitude (see Eq. (1.27)).

The above elimination of intermediate states can also be performed in a straightforward way for other multiphoton processes for the general case of off-resonant intermediate states reducing the equation of motion again to the form of a two-level system. In case of an n-photon transition, the resulting n-photon Rabi frequency, $\chi_n(t)$ will then be proportional to the nth power of the field amplitude and the transition probability between the initial and final states will then be proportional to the nth power of the field intensity for low intensities.

The population dynamics for a TPA is demonstrated in Fig. 3 in case of a single intermediate state. Here, the original TDSE was solved and the population, P_m of the intermediate state is displayed together with the ground- and excited state populations, P_g and P_e, respectively. (The two-photon detuning was set to zero, the energy of the various states as well as the peak field strength are given in the figure.) Panels (a) and (b) show the time evolution of populations for a small detuning in case of moderate and high field strengths, respectively. It is seen that the intermediate state is little populated throughout the process, despite the relatively small detuning even in the case when Rabi oscillations between states g and e occur. It can be seen in panel (e) that for low-field strengths, the excited-state population is proportional to the fourth power of the field strength, in accordance with Eq. (1.27). Panel (f) shows that apart from a small range of detuning at around ($V_m - V_0 \approx \hbar\omega_0$), where net population is transferred also onto the intermediate state, the conditions for the adiabatic elimination procedure are fulfilled. In the special case of a full Rabi cycle, the excited net population is zero (see Fig. 3(d)), while in case of half a Rabi cycle, complete population transfer can be achieved even for large detuning.

Another well-known two-photon process, where complete population transfer can be achieved while the intermediate state is not populated throughout the process, is the Stimulated Raman Adiabatic Passage (STIRAP).[8] In STIRAP, the population transfer between states (g) and (e) is mediated by a state, m, lying higher in energy than both the (e) and (g) states and the transition from g to e is achieved by two subsequent laser pulses of different wavelengths, the first one being tuned to ω_{em} and the second one being tuned to ω_{mg}.

1.2.6. *Laser-dressed states*

In Sec. 1.2.4, we have defined how the energy of two levels can be shifted by an electric field. In this section, we will extend the Stark-shift equations to molecular potentials, where the energy depends on the geometry of the molecule, i.e., the shape of the potentials may play a very important role. Figure 4(a) shows the typical behavior of the electronic potential energy in a one-dimensional potential for two bound potentials V_m and V_n, i.e., both curves have a minimum located at different R, respectively. As in the case of the two-level system (Sec. 1.2.4), we can use the interaction picture to include the laser-photon energy in the diagonal part of the potential

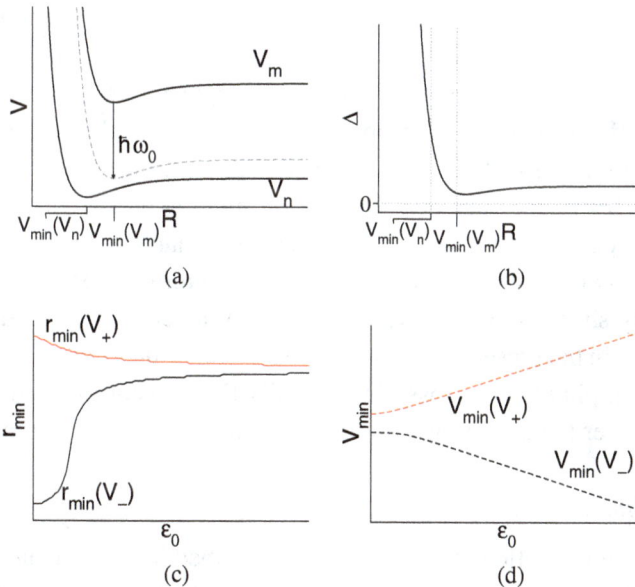

Fig. 4. Laser-dressing of molecular potentials: (a) Bare potentials, where V_m is represented as the gray, dashed line in the interaction picture. (b) The detuning Δ is dependent on the coordinate. (c) Location of the potential minimum of the dressed states ($r_{min}(V_{\{+,-\}})$) vs. field strength. (d) Energy of potential minimum of the dressed states ($V_{min}(V_{\{+,-\}})$) vs. field strength.

interaction matrix:

$$\underline{\underline{W}}_I(R, t) = \begin{pmatrix} V_n(R) & -\dfrac{\mu_{nm}(R)\varepsilon_{\mathrm{env}}(t)}{2} \\ -\dfrac{\mu_{mn}(R)\varepsilon_{\mathrm{env}}(t)}{2} & V_m(R) - \hbar\omega \end{pmatrix}, \qquad (1.29)$$

where the detuning of the laser with respect to the transition is $\Delta(R) = V_m(R) - \hbar\omega - V_n(R)$, see Fig. 4(b). Since the shapes of the electronic potentials m and n are different, the detuning changes with the R coordinate, and the Stark effect is not the same for every R. As a consequence, the potentials are not only shifted but reshaped under the effect of the electric field creating a new set of potentials, the so-called Light Induced Potentials (LIPs) or dressed states. These can be calculated by diagonalizing the matrix $\underline{\underline{W}}_I$,

$$\underline{\underline{W}}_I = \underline{\underline{Z}}^\dagger \underline{\underline{D}}_I \underline{\underline{Z}}, \qquad (1.30)$$

where $\underline{\underline{D}}_I$ is the diagonal matrix containing the energies of the new LIPs (denoted V_+ and V_- here) and $\underline{\underline{Z}}$ represents the composition of these new electronic potentials in the original, bare state, picture.

The effects on the LIPs' properties for different field amplitudes ε_0 are shown in Fig. 4(c), 4(d). On the one hand, we can see the repulsion between the two LIPs as the energetic shift of the minima of V_- and V_+, similar to the one described in Sec. 1.2.4. On the other hand, the reshaping of the potentials is observed as the change of the minimum location, which is drastically shifted towards large distances in the case of V_- and to lower ones in V_+ in the present example. Moreover, since the position depends on the laser amplitude, it is possible to modify the new equilibrium geometry with the laser field, opening new strategies to control the dynamics.

1.2.7. *Photon locking*

In the previous section (1.2.6), we have described how strong nonresonant laser fields are able to modify the electronic potentials. The change of the electronic-states properties due to this modification can be used to create new control schemes, for example to trap a molecule in a specific geometry. Several studies on spatially trapping a molecular wavefunction exist, see, e.g., Ref. 73. A compelling approach was introduced by Sola and coworkers in the Laser Adiabatic Manipulation of the Bond (LAMB) control scheme,[74] where the equilibrium is modified by creating a LIP. In this scheme, this LIP is adiabatically created and the wavepacket is always a vibrational eigenfunction during the dynamics.

In contrast to the LAMB method, in the photon locking scheme, the control is achieved over a nonstationary vibrational state that is previously promoted to a bare electronic state, for example by ionizing the molecule. During the dynamics, the control laser creates a barrier that reflects the wavepacket and restricts its movement. The key of this scheme is the frequency of the laser field that is chosen to put V_1 and V_2 into resonance at an intermediate geometry between the promoted wavepacket and the minimum of V_1, as depicted in Fig. 5.

In this way, after the wavepacket is promoted to V_1, it moves in the direction of the minimum of V_1, i.e., from right to left. Before the wavepacket arrives to the potential minimum, the control laser field is applied, creating a series of LIPs (V_- and V_+). Since the wavepacket has not yet reached the Franck–Condon region of the control laser field, it stays in V_-. When arriving at the Franck–Condon region, the steep slope of V_- blocks the way of the wavepacket. Finally, the wavepacket is reflected back and, if the creation of the LIPs is adiabatic, there is no excitation to V_2.

1.2.8. *Hole burning*

In common hole burning, the absorption spectrum of a molecule exhibits a "hole" at a certain frequency because the considered molecule is changed or destroyed by the interaction with a light of this frequency.[75] Usually, continuous-wave lasers are applied to molecules in their ground state to achieve hole burning.

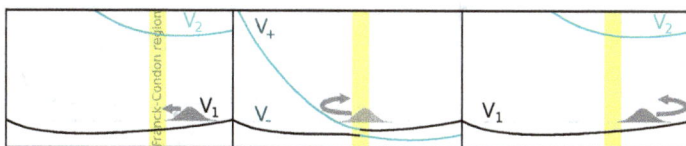

Fig. 5. Photon locking scheme. A wavepacket is created on V_1 close to the Franck–Condon region of a control laser (left panel). The control laser couples V_1 and V_2, creating the LIPs V_- and V_+, and traps the wavepacket (middle panel). After the laser interaction, the wavepacket can again evolve freely on V_1 (right panel).

In what we term hole burning,[24] a somewhat different situation is described. We look at nuclear wavepackets instead of spectra and the hole is situated in coordinate space instead of in the frequency domain. The hole in the wavepacket is created by an ultrashort laser pulse instead of a continuous-wave laser. This short pulse is able to do hole burning if its Franck–Condon region is smaller than the width of the wavepacket and the pulse duration is short compared to the velocity of the wavepacket. In this case, the fast and sharp laser is "perforating" the big and slow wavepacket. This behavior can be rationalized in a dressed-state picture (see also Sec. 1.2.6).

In Fig. 6, the situation before (left panel), during (middle panel), and after (right panel) the laser interaction is sketched, respectively. Before the laser is turned on, a wavepacket moves on a potential, which we term V_1. When the laser is acting, the potentials are mixed resulting in a V_+ and a V_-. The wavepacket moves on V_-, which has contributions from V_1 and V_2, tentatively indicated by the color coding. If the laser intensity is very strong, the wavepacket will remain on V_-, moving from the black region to the cyan region, which means a population transfer to V_2. If the laser pulse starts interacting with the molecule, while the wavepacket is already in the Franck–Condon region (close to where the colors are interchanged in Fig. 6 (middle panel)) and ends before the wavepacket completely leaves this region, then only a part of the wavepacket is transferred to V_2. The remaining wavepacket in V_1 exhibits a hole.

The effect of this type of control is that the wavepacket shape in coordinate space is changed. In the discussed case, the width is diminished. As coordinate and momentum are related via a Fourier transform, also

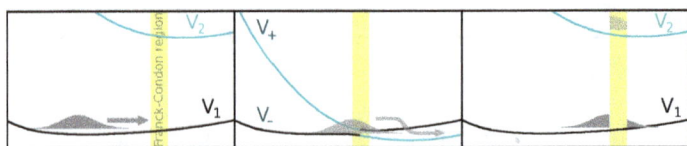

Fig. 6.   Hole burning scheme. A comparably slow and widespread wavepacket moves towards a future, localized Franck–Condon region (left panel). The laser interaction can be understood in the field-dressed picture with potentials $V_{\{+,-\}}$, where the wavepacket mainly moves on V_- (middle panel). After the laser pulse is over, the original potentials $V_{\{1,2\}}$ are restored and a hole is created in the wavepacket in V_1 (right panel).

the momentum distribution will change. In the above case, the momentum distribution will be wider. Such effects can be observed experimentally.

1.2.9. *Strong-field ionization*

While the interaction between a molecule and a weak electromagnetic field can be described using perturbation theory, strong-field molecular ionization is a complicated nonperturbative multi-electron process. For such processes, there are currently no complete theories which are able to predict molecular-ionization yields, even given a fairly good understanding of the molecular structure. Historically, descriptions of strong-field molecular ionization have drawn upon ideas from strong-field atomic ionization, for which simple and intuitive models have been developed.[76,77] Figure 7 illustrates the distortion of the atomic binding potential under the influence of a strong electric field at the peak of an oscillating laser pulse. Of course, the potential for a molecule is more complicated, but for illustrative purposes we limit ourselves here to a discussion of a simple unstructured Coulomb potential for a single atom.

There are two important regimes which are relevant for laser-driven SFI. One is the so called "multiphoton" regime, which corresponds to the case where the ionization takes place over many cycles of the laser field and in which case the ionization rate is much less than the laser frequency.

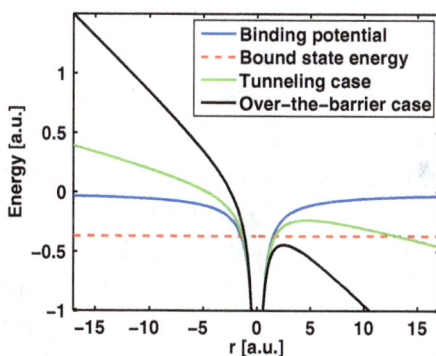

Fig. 7. (color online) Binding potential of an atom (blue solid line), valence electron energy based on a 10 eV ionization potential (red, dashed line), laser-dressed binding potential in the tunnel-ionization case (green solid line), laser-dressed binding potential in the over-the-barrier case (black solid line).

The complementary regime known as the "tunnel" regime corresponds to the case where the laser frequency is low in comparison to the ionization rate, and therefore significant ionization can take place in a half cycle of the laser field. These two regimes are typically distinguished quantitatively by the Keldysh adiabaticity parameter γ:[78]

$$\gamma = \sqrt{\frac{I_P}{2U_P}}, \quad U_P = \frac{|\varepsilon_0|^2}{4\omega_{laser}^2}. \tag{1.31}$$

Here, I_P is the ionization potential, U_P is the ponderomotive energy, or the average energy of electron oscillations in the laser field, ω_{laser} is the laser frequency and $|\varepsilon_0|$ the electric field amplitude. Quasi-static tunneling corresponds to $\gamma \ll 1$, while multiphoton ionization corresponds to $\gamma \gg 1$. It is useful to define the concept of tunneling time, which is the time it would take for the electron to cross the barrier moving in a uniform electric field, if the process were classically allowed. For this process (setting electron mass, $m_e = 1$, and electron charge, $e = 1$), the velocity of the electron as a function of time is given by $v(t) = v_{max} - |\varepsilon_0|t$. Here $v_{max} = \sqrt{2I_P}$, and for tunneling resulting in an electron produced in the continuum with zero energy, $v_{final} = 0$. This yields a tunneling time of $\tau_{tunnel} = v_{max}/|\varepsilon_0| = \sqrt{2I_P}/|\varepsilon_0|$. Expressing the Keldysh parameter in terms of the laser frequency ω_{laser}, and the tunneling frequency, defined as $\omega_{tunnel} = 1/\tau_{tunnel}$ leads to the expression:

$$\gamma = \frac{\omega_{laser}}{\omega_{tunnel}}. \tag{1.32}$$

The quasi-static tunneling regime is characterized by $\omega_{laser}/\omega_{tunnel} \ll 1$. In this limit, the shape of the barrier does not change significantly during the tunneling process,[79] giving rise to the name used for this regime. It is an interplay of laser frequency, ionization potential and field strength that leads to quasi-static tunneling being the dominant effect in an ionization process. The field strength has to be high enough to tilt the potential sufficiently to give rise to a finite barrier, while the frequency has to be low enough that the condition from Eq. (1.32) is satisfied for tunneling to take place on a subcycle timescale. It should be noted that a tunneling component is present in the multiphoton regime as well. This tunneling differs from the quasi-static one, in that the barrier shape changes during the tunneling process.

At very high intensities, the electric field of the laser can tilt the Coulomb potential and completely suppress the barrier to ionization, making the electron escape classically allowed. The observed intensities of appearance of several charge species of noble gases agree well with the prediction of the simple, semi-classical model.[80]

Theoretical efforts in understanding the tunneling process started with the development of quantum mechanics. A common feature of these theories is that the ionization rates depend strongly (exponentially) on the binding potential. Tunneling theory was first derived by Fowler and Nordheim[81] in 1928, for the case of electron emission from metals. Oppenheimer[82, 83] applied it to ionization of hydrogen-like atoms in strong external fields. It was later rederived by Keldysh[78] and by Perelomov, Popov and Terentev,[84] for DC tunneling from hydrogen-like atoms in a field of a strong oscillating electromagnetic field. A treatment of non-hydrogen, polyelectron atoms was presented by Ammosov, Delone and Krainov in 1986,[85] and became known as the ADK tunneling theory. A further level of sophistification was added by Faisal and Reiss,[78, 86–89] in what is known as the Keldysh–Faisal–Reiss (KFR) theory. This is closely related to the strong-field approximation (SFA), which has become a standard approach to calculating SFI yields.

The SFA calculates the ionization amplitudes with an S-matrix formalism and treats the continuum states as solutions to free electrons oscillating in the laser field alone ignoring the effects of the ionic electrostatic potential on the continuum states. This treatment of the continuum is qualitatively similar to the first Born approximation of scattering states with the addition that the action of the laser is taken into full account. Allowing the laser field to act only on one electron (the single-active electron approximation, almost universally invoked in SFA treatments of strong-field effects), the SFA probability for ionization of a multielectron target, with corresponding generation of a continuum electron with momentum $\mathbf{k}$, can be written as

$$w_{lm}^{\text{SFA}}(\mathbf{k}) = |\langle \phi_l^{\mathbf{k}} | U^{\text{SFA}} | \tilde{\phi}_{lm}^D \rangle|^2 G_{lm}^D, \qquad (1.33)$$

where G_{lm}^D is the Dyson norm, calculated from an ionic state $|I_l\rangle$ with quantum number l and a neutral one $|N_m\rangle$. For the Dyson norm, unnormalized Dyson orbitals are used while the $\left|\tilde{\phi}_{lm}^{D}\right\rangle$ refers to a normalized Dyson orbital. The $\left\langle\phi_{l}^{\mathbf{k}}\right|$ denotes the final state of the ionization process, i.e., consists of an ionic state and a continuum electron. U^{SFA} is the SFA propagator and is given by

$$U^{\text{SFA}} = -i \int dt' \int d\mathbf{k} \left|\phi_{l}^{\mathbf{k}}\right\rangle$$
$$\times \exp\left[-\frac{i}{2}\int_{t'}^{t} |\mathbf{k} + \mathbf{A}(\tau)|^{2} d\tau\right] e^{iI_{p,lm}t'}\left\langle\phi_{l}^{\mathbf{k}}\right|V(t'), \quad (1.34)$$

where $I_{p,lm} = E_{I}^{l} - E_{N}^{m}$ is the ionization potential (I_{p}), $\mathbf{A}(\tau)$ is the vector potential and $V(t) = \sum_{j=1}^{n} \varepsilon(t)\boldsymbol{\epsilon}_{\varepsilon} \cdot \mathbf{r}_{j}$ with $\varepsilon(t)$ being the time-dependent electric field of the laser and $\boldsymbol{\epsilon}_{\varepsilon}$ the polarization direction. The integral in Eq. (1.34) can be solved using approaches based on stationary phase and/or semiclassical approximations.[78,84,85] Following integration, the total yield can be written as

$$W_{lm}^{\text{SFA}} = C_{lm} K(I_{p,lm}, \varepsilon_{0})G_{lm}^{D}, \quad (1.35)$$

where

$$K(I_{p,lm}, F_{0}) = \exp\left[-\frac{2}{3}\frac{(2I_{p,lm})^{3/2}}{|\varepsilon_{0}|}\right], \quad (1.36)$$

is the dominant exponential factor of the Keldysh tunnel-ionization rate,[78] and C_{lm} is a prefactor that depends weakly (i.e., not exponentially) on the field strength ε_{0} and $I_{p,lm}$ and also depends on the specific state (or Dyson orbital) being ionized.

For molecular systems with low lying electronic states of the molecular cation, ionization to excited ionic states can compete with ionization to the cationic ground state. The SFA predicts that for molecules with ionization potentials of *ca.* 10 eV, and laser intensities at which ionization to the ground state becomes appreciable (10^{13} W/cm^{2}), ionization to excited states of the molecule can be non-negligible.[90] Furthermore, the recently developed time-dependent resolution in ionic states (TDRIS) approach to calculating SFI yields, which goes beyond the SFA, gives excited-state yields in excess of the SFA predictions.[91] Experimental measurements are in agreement with the TDRIS calculations, indicating that SFI of molecules with low

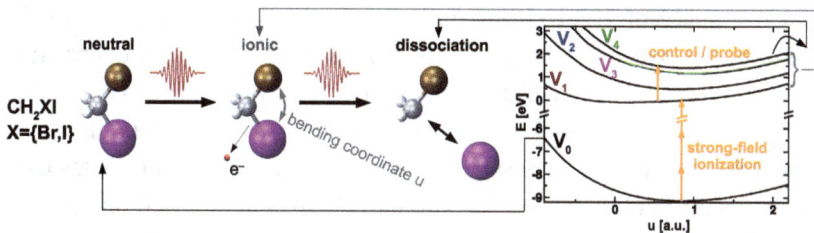

Fig. 8.   Excitation scheme.

lying, closely spaced cationic states can lead to superpositions of ionic states (multi-hole electronic wavepackets).[90] The latter will be discussed in the context of halogenated methane molecules below.

1.3. Computational and experimental details

In the examples presented in the sections below, we consider the halogenated methanes CH_2BrI and CH_2I_2. They are first multiphoton-ionized by a strong-field pump pulse. The subsequent dynamics of the ions is then controlled and/or probed with further pulses, which may lead to dissociation of the ionic compounds as seen in Fig. 8.

In order to understand the experimental results, electron-structure calculations are carried out to obtain potential-energy curves, whereupon quantum-dynamics is simulated. For the electronic-structure calculations (see also Sec. 1.2.1), we rely on three methods. DFT with the B3LYP functional[92] and the aug-cc-pVTZ basis set extended with effective core potential[93] for the iodine atom within the Gaussian03 program[94] are used to optimize geometries and determine normal vibrational modes in ground electronic states. The most important coordinate in the considered systems is the I–C–Br or I–C–I bending coordinate, respectively, termed u as seen in Fig. 8. Consequently, the calculations are restricted to this one degree of freedom. Note that such a simplification is supported by the good agreement with experimental measurements, as demonstrated in the following sections and previous publications.[24,95–98] We calculate the potential energy curves for the neutral ground state V_0, the ionic ground state V_1 and ionic excited states (V_2 - V_5 for both CH_2BrI and CH_2I_2) as well as the corresponding TDM and SOC curves with the SA-CASSCF method. Here, we employ an active space of 12 electrons for the neutral or 11 electrons for the ion in 8 orbitals. We used the ANO-RCC basis sets[99, 100] with contractions equivalent to a triple-zeta basis with polarizations (3s2p1d for H, 4s3p2d1f for C, and 6s5p3d2f1g for Br as well as I atoms). The Douglas–Kroll Hamiltonian is applied and SOC is computed among the ionic states using atomic mean-field integrals.[61] These calculations are carried out with the MOLCAS 7.2 package.[101]

For the quantum dynamics calculations, we solve the TDSE (cf. Sec. 1.2.3) employing these potentials. As the ionic states are doublets, two degenerate potential curves represent each ionic state, where the latter are nonetheless termed V_1, V_2, etc., respectively (Fig. 8). In this case, each component m of the wavefunction with $m > 0$ and each element mn of the potential- and TDM matrices with $m, n > 0$ in Eq. (1.7) should be considered as

$$
\psi_m^{di} = \begin{bmatrix} \psi_m^{di,+} \\ \psi_m^{di,-} \end{bmatrix}, \quad V_{mn}^{di} = \begin{bmatrix} V_{mn}^{di,++}, & V_{mn}^{di,+-} \\ V_{mn}^{di,-+}, & V_{mn}^{di,--} \end{bmatrix},
$$

$$
\mu_{mn}^{di} = \begin{bmatrix} \mu_{mn}^{di,++}, & 0 \\ 0, & \mu_{mn}^{di,--} \end{bmatrix}. \tag{1.37}
$$

Here, we add the index di to indicate that we work in a diabatic representation and the superscript $\{+, -\}$ relate to the different spin states (see also Sec. 1.2.2).

The electric field of the pump and the control pulses are included explicitly in the TDSE. However, the SFI process is modeled qualitatively using a UV (ultraviolet) pulse tuned to resonance between V_0 and V_1 instead of the true strong IR (infrared) pump pulse inducing multiphoton processes. Note however that it is possible to account for the dynamic Stark effect due to the strong IR pump pulse by modeling the pump pulse by two electric fields, for details see Ref. 24. The TDSE is solved on a grid of 128 points using the SO technique.[67] A time-step of 0.01 fs and a spatial discretization of 0.025 a.u. is employed. The grid size was checked for convergence. A detailed description of the complete computational procedure may be found in Ref. 95.

Our experimental measurements make use of pulses from an amplified titanium:sapphire laser system with a minimum pulse duration of 30 fs and a central wavelength of 780 nm. It can deliver pulses of 1 mJ at a repetition rate of 1 KHz that are subsequently split into pump and probe pulses in a Mach–Zehnder interferometer. We perform measurements for both IR and UV pump and probe pulses. The UV pulses are generated via third-harmonic generation of the laser output to produce pulses with a central wavelength of 260 nm. The "pump" arm contains a pulse shaper with a computer-controlled acousto-optic modulator (AOM) as the shaping element,[102] allowing us to change the spectral phase, energy and delay of the pump pulses. The probe arm of the interferometer contains a delay stage for mechanical delay of the probe pulses relative to the pump. The pump and probe pulses are focused and intersect in an effusive molecular beam inside a vacuum chamber equipped with a spectrometer that can be operated in time-of-flight or velocity-map-imaging (VMI) mode. Focused intensities of the pump and probe pulses are varied from 1×10^{12} W/cm^2 to 1×10^{14} W/cm^2.

1.4. Vibrational Wavepackets Created by Multiphoton Ionization

1.4.1. *Phase-dependent dissociation*

Joint experimental and theoretical investigations on dihalomethane molecules have made it possible to demonstrate various phenomena related to strong-field excitations. One of these is the phase-dependent dissociation, observed for CH_2BrI^+ in pump-probe measurements. In this case a strong laser pulse excites a propagating wavepacket to a dissociative electronic state, where the transition probability depends on the momentum — i.e., on the spatially varying phase — of the wavepacket rather than on the position of the wavepacket. The process is explained schematically in Figs. 9(a) and 9(b), Depending on the direction of wavepacket propagation on the lower state (V_l) relative to the slopes of the potentials, the laser field does or does not have the chance to de-excite the previously excited molecule: In case (a), the excited wavepacket slows down on the excited-state potential (V_u) and returns back to the resonance region. Thus, it can be de-excited by the same pulse to the initial state, while in case (b) the wavepacket does not return to the resonance location, once it is excited by the laser field. Consequently, the final population in the upper state (i.e., the

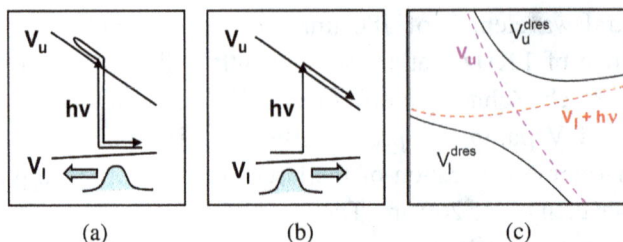

Fig. 9.   Schematic description of momentum dependent transition. Depending on the propagation direction and the potential slopes, de-excitation after an excitation is more (panel a) or less (panel b) favored. For strong laser fields, the dressed states have to be considered (panel c).

number of excited molecules) is different in the two cases. Process (a) is a multiphoton process requiring relatively strong laser fields. If, however, the laser is considerably strong, then the wavepacket dynamics is governed by the upper or lower dressed-state potential (see Fig. 9(c)), depending on its initial position and momentum on the lower state V_l before the laser is turned on.

In the experiments on CH_2BrI^+, a 40 fs strong IR pump pulse of 784 nm is used to ionize the molecules and a subsequent IR pulse of the same wavelength and duration is used to probe the dynamics induced by the pump pulse. Although the probe pulse is not strong enough to ionize the neutral molecules due to the large difference between the photon energy (1.58 eV) and the ionization potential (9.69 eV) of the molecule, its intensity is enough to induce multiphoton transitions within the parent ion created by the pump pulse. As a result of the interaction with these laser fields, some of the molecules dissociate in the ionic continuum and the fragment ions as well as the parent ions are detected by time-of-flight mass spectrometry (TOFMS). The main products detected are the parent ion and CH_2Br^+. Their normalized TOFMS signals as a function of pump-probe delay time are shown in Fig. 10.

These signals show complementary oscillations (i.e., perfectly π out of phase) for positive delay times. The most important property of the parent (fragment) ion signal is that the periodic modulations consist of subsequent small and large dips (hills) indicated, respectively by solid red and dashed black arrows in the figure. As a consequence of this

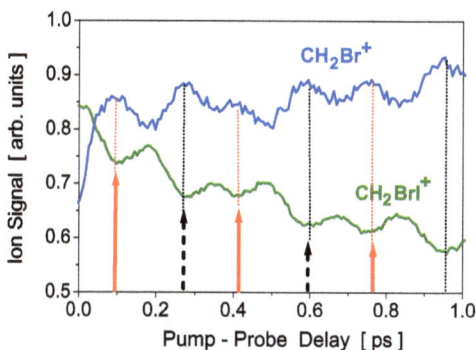

Fig. 10. (color online) Pump-probe signal measured by TOFMS. The parent-ion (CH_2BrI^+) and fragment-ion (CH_2Br^+) yields exhibit complementary oscillations as indicated by the red and black arrows.

modulation structure, the Fourier transform of the signal yields two frequencies: The fundamental at $94 \pm 4\,cm^{-1}$ and its second harmonic one at $196 \pm 4\,cm^{-1}$. The fundamental frequency matches very well the Br–C–I bending frequency of the parent ion at FC_{pump}. Furthermore, the equilibrium Br–C–I angles in the neutral and ionic electronic ground states are known to differ considerably (ca. 20°) while the rest of the geometry parameters are more or less the same. In addition, it turns out from single photon dissociative photoionization measurements,[103] that electronic states beyond the first excited state should be — directly or indirectly — dissociative. All these issues together suggest that the observed oscillations in the pump-probe signal are due to a Br–C–I bending motion induced by the pump pulse in the ground electronic state of the parent ion.

In order to model the process, we first determine one-dimensional potential-energy curves for the neutral ground electronic state and for the lowest five ionic electronic states along the bending normal mode of the cation as described in Sec. 1.3. Diabatic and adiabatic potentials — the latter including the effect of SOC — for the cation are presented in Fig. 11.

Having the potentials for the cation and for the ground state of the neutral molecule, we performed quantum-dynamics simulations, where we approximated the multiphoton ionization with a single-photon excitation using reduced intensity. Although ionization occurs to several electronic states — as will be discussed in later sections — the periodic modulations

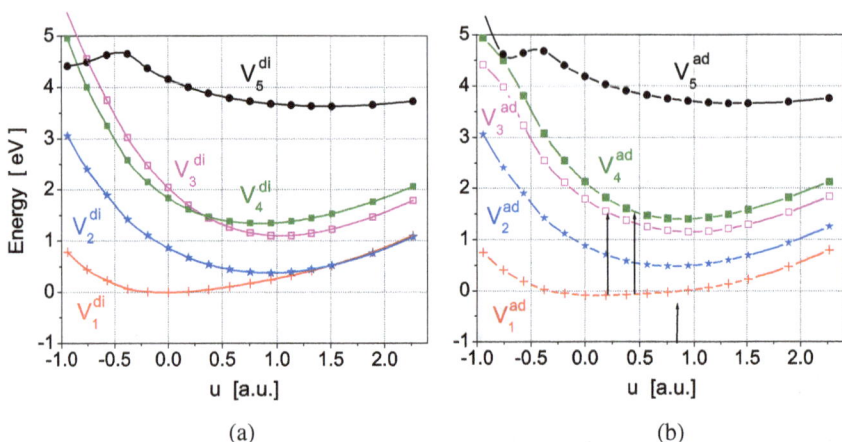

Fig. 11.   Potential energy curves for the CH_2BrI^+: Panels (a) and (b) show the lowest five diabatic and adiabatic potentials of the ion, respectively.

are the result of vibrational wavepacket propagation in the ground ionic state. Therefore, we adjusted the wavelength of the pump pulse to the ionization potential of the molecule, focusing only on the periodic modulations in the ion signal. For the probe pulse, we used the true experimental pulse parameters. Since the excited states V_3^{ad}, V_4^{ad} and V_5^{ad} (where the index ad indicates that SOC has been included in these curves, see also Secs. 1.3 and 1.2.2) are either directly or indirectly dissociative, the fragment ion signal is related to the sum of the final populations of these states excited by the probe pulse from the ground ionic state, V_1^{ad}.

The results of the dynamical simulations explain the modulation structure in the experimental signal. The pump pulse launches a vibrational wavepacket in the ionic ground state V_1^{ad} that moves towards smaller I–C–Br bending angles (smaller normal mode coordinate values, u, see Fig. 11). During its coherent back-and-forth motion on V_1^{ad}, the wavepacket crosses a position, (indicated by the "probe" arrows in Fig. 11b) at around $u = 0.34$ a.u. twice in each vibrational period, at which it can be resonantly excited to the upper potentials V_3^{ad} and V_4^{ad}. (We call this position the Franck–Condon position for the probe: FC_{probe}.) When the pump-probe delay time matches the time at which the wavepacket is at around FC_{probe}, the excitation becomes very efficient, resulting in increased populations in states 3 and 4. This, in turn, is reflected as peaks in the observed fragment

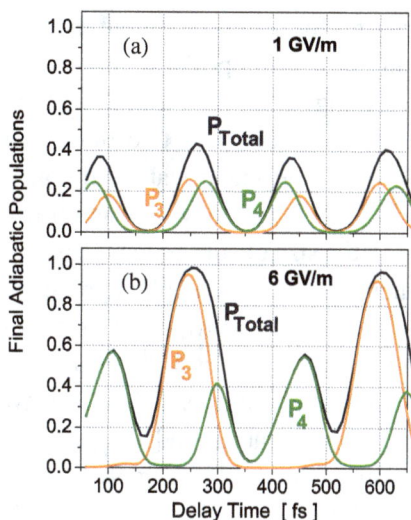

Fig. 12. (color online)Populations of excited states: Panels (a) and (b) show the final populations in adiabatic excited states for peak probe field strength of 1 and 6 GV/m, respectively. Populations in other excited states are negligible and therefore not displayed. Populations of individual adiabatic states 3 and 4 and total excited populations P_{Total} are shown by orange, green and dark grey curves, respectively.

ion signal. Population values in states 3 and 4, obtained from the numerical simulations, are shown in Fig. 12 as a function of pump-probe delay. The positions of the P_{Total} peaks perfectly match those of the CH_2Br^+ signal in Fig. 10. The subsequent occurrence of small and high peaks is also reproduced. The reason for this feature is exactly the same as schematically explained by Fig. 9: The momentum of the ground ionic state wavepacket is opposite when comparing the first and the second crossing of the resonance location within one vibrational period. When moving towards larger u values (opening the I–C–Br angle), the probe pulse can only excite the molecule once, as indicated schematically in Fig. 9b, while in the other case, i.e., when the wavepacket moves towards smaller u values (closing I–C–Br angle), the probe pulse has the chance to de-excite the wavepacket from V_3^{ad} and V_4^{ad} according to Fig. 9a, leaving less final population in states V_3^{ad} or V_4^{ad}. This second process is, however, a multiphoton one and thus requires high intensity. With low intensities such as, e.g., at 1 GV/m peak field strength, the probability of de-excitation after excitation is negligible and the net population transfer for the two different momenta is approximately the same resulting in subsequent peaks of almost the same size in the pump-probe signal (see Fig. 12a). Increasing the intensity, the difference between the size of the peaks belonging to different wavepacket momenta is increasing, as shown by the numerical results in Fig. 12b for a peak field strength of 6 GV/ m. This behavior was observed experimentally: The difference between the size of subsequent peaks (dips) in the fragment (parent) ion signal increased with the applied probe laser intensity.

The fine details, such as the distribution of final populations among the close lying excited adiabatic states V_3^{ad} and V_4^{ad} can be understood by considering, that in strong enough laser fields the wavepacket follows the dressed-state potentials shown at around FC_{probe} in Fig. 13. Unlike the simple schematic situation indicated by Fig. 9c in our particular case there are two adiabatic excited states strongly coupled to V_1^{ad}, and thus two upper dressed states, V_3^{dres} and V_4^{dres}, which asymptotically correlate with V_4^{ad} and V_1^{ad} for u-values larger than FC_{probe}. Since the wavepacket moving towards smaller u-values on V_1^{ad} in the $u > FC_{probe}$ region is broad enough, both upper dressed states will be populated when the laser is turned on. The corresponding two wavepackets will be reflected by the wall of these dressed potentials and end up finally in the correlated adiabatic states V_1^{ad} and V_4^{ad}, but not on V_3^{ad} (see the populations in Fig. 12b at around delay times 100 fs and 450 fs). In contrast, when the wavepacket moves initially towards larger u values in the $u < FC_{probe}$ region of V_1^{ad} and follows the

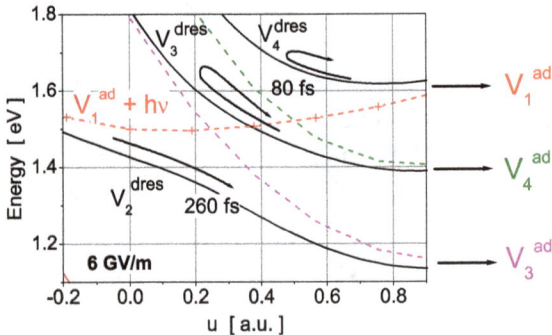

Fig. 13.   Dressed-state picture. Solid black lines show the important dressed-state potentials for 6 GV/m field strength. The field-free adiabatic potentials are also displayed by colored dashed lines.

lower dressed potential, it will exclusively end up in the correlated adiabatic state V_3^{ad}, as sketched in Fig. 13.

1.4.1.1. *Photon locking*

In the following, we discuss an application of photon-locking control (recall Sec. 1.2.7) in the CH_2BrI molecule.[24] First, the molecule is ionized via a multiphoton process and we obtain CH_2BrI^+ as indicated in Fig. 8 and as discussed also in the previous section. Thus, a wavepacket on the ionic ground-state potential V_1 is created at FC_{pump}. It oscillates on V_1 with a period of 351 fs primarily along the I–C–Br bending coordinate u (see Sec. 1.3), if no control laser is interacting. The dynamics is probed with a UV pulse (260 nm), whereby population is transferred to a very high-lying state, that we term V_n, at FC_{probe} (not the same as in previous section), very close to FC_{pump}. From this V_n, CH_2BrI^+ dissociates yielding CH_2I^+ and thus, the wavepacket motion is mapped to the time-dependent CH_2BrI^+ and CH_2I^+ signals in a complementary way.

The control laser (780 nm) is able to couple V_1 to the ionic states V_3, V_4 and a dressed-state potential V_+ is obtained, as detailed in Fig. 5. The latter potential changes its character from that of bare V_1 to bare $V_{3/4}$ at $FC_{control}$ (being the same as FC_{probe} of the previous section).[95] The slope of $V_{3/4}$ is much higher than the one of V_1 at $FC_{control}$, so that — comparing the original V_1 and the new V_+ — the situation can be pictured as creating a "wall", which can block the wavepacket from propagating. In the experiment, this effect is obtained by superimposing the ionizing pump pulse and the control pulse. The latter is stretched in time by adding second-order phase, where energy and duration are modified at our pulse shaper, between 0 and $19.6 \mu J$, and 75 fs and 240 fs (FWHM) respectively. In this parameter range, no ionization due to the control pulse alone is observed.

If the control pulse field strength is high enough, the "wall" will be "solid" enough to trap the wavepacket on V_+, where the wavepacket will oscillate with much smaller amplitude compared to V_1. When looking at the potential shape of V_+ (see Fig. 5), it becomes clear that the wavepacket will spend most of the time in regions corresponding to V_1. Hence, the population transfer to $V_{3/4}$ is negligible, as confirmed by both experiment and theory.

When switching off the control pulse, we can distinguish two scenarios. If the control pulse is turned off adiabatically, i.e., slowly compared to the oscillation period, then kinetic energy will be lost to the control field. As a consequence, the oscillation amplitude will decrease and FC_{probe} will not be reached by the complete wavepacket anymore. In this case, a decreased modulation depth in the parent ion signal is expected.

In contrast, if the control pulse is switched off impulsively, i.e., fast compared to the oscillation period, the oscillation amplitude will be retained compared to the one without control. However, a delay of the oscillations proportional to the length of the control pulse duration will be observed.

In order to analyze the two expected effects — the decreased modulation depth and the delayed oscillations — in an easily discernible fashion, we performed Fourier transforms of the respective pump-probe signal for various control-pulse durations.[24] The amplitude of the Fourier transform at the V_1 frequency (corresponding to a period of 351 fs) quantifies the amplitude of the wavepacket on V_1. Hence, the modulation depth can be directly monitored. The phase of the Fourier transform is proportional to the displacement of the wavepacket on V_1 and thus, yields information on a delay of the wavepacket oscillations.

Such a delay is then, in principle, visible as a shift of the phase. In order to witness the phase shift, all molecules in a sample should experience the same field strength. However, this is not the case due to the orientational distribution of the molecules. Accordingly, we are not able to distinguish a considerable phase shift in our signals.

Due to our above considerations for the adiabatic control pulse switch-off, we show the modulation depth in both the calculated and experimental pump-probe signal in Fig. 14. As expected, the modulation depth decreases with increasing intensity and time duration of the control pulse in both cases. Small deviations between experiment and theory can be explained with angle averaging effects naturally occurring in the experiment, but not included in the simulations.

1.4.1.2. *Hole burning*

In this section, we discuss an application of the hole-burning control scheme (see Sec. 1.2.8) using the CH_2BrI molecule as a test system and the same setup as in the previous section.[24] Hole burning can be observed in the

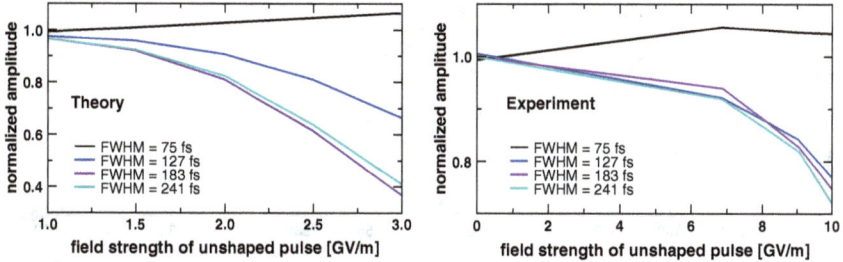

Fig. 14. Fourier transform amplitude at the V_1 bending frequency for pulses of different duration (obtained by second-order chirp). The amplitude decreases for stronger and/or longer fields, indicating a decrease of the modulation depth of the parent ion yield, as expected for photon locking. Theory (left panel) and experiment (right panel) qualitatively agree. In the simulations, the increase in modulation depth with field strength for the shortest control pulse duration is an artifact of the simple model used to describe the ionization process.

present case, everytime when the wavepacket moves from the inner to the outer turning point on V_1, i.e., at the second/fourth/etc passage through $FC_{control}$ (see Fig. 6).

In the experiment, the fourth passage (centered at 610 fs) has been chosen for the analysis of the recorded data, corresponding to pump-control time delays between 560–660 fs. Such a long delay can still be used to demonstrate the hole-burning, since the wavepacket shows little de-phasing within the first picoseconds. The probe-pulse delay was taken to be in the interval between 1.1 ps and 2.2 ps in order to avoid optical interference.

As described in the previous section, the analysis is facilitated by employing Fourier transforms of the pump-probe signal. The so-obtained amplitude corresponds to the wavepacket amplitude and the phase indicates a shift of the wavepacket's center of gravity. The results are shown for both measurement and simulation in Fig. 15.

We observe a decrease in amplitude when the control-delay time matches with the wavepacket on V_1 being in the $FC_{control}$ region. Due to the strong control pulse, a significant amount of population is transferred to $V_{3/4}$. As a consequence, the wavepacket remaining on V_1 changes its shape. The fact that we observe hole burning indeed, can be concluded from the measured change of the phase in Fig. 15 (right panel). A localized fraction of the wavepacket is removed by the control pulse, which leads to a shift in the wavepacket's center of gravity and creates additional momentum

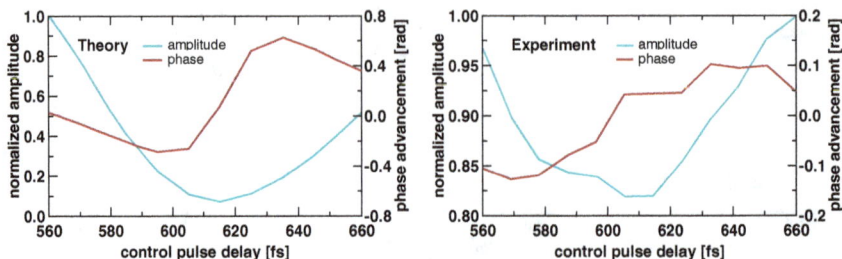

Fig. 15.　Amplitude and phase advancement of the Fourier transform of the pump-probe signal at the V_1 vibrational frequency vs. control-pulse delay time. The phase advancement is the difference in phase of the pump-probe signal for the evolution with and without the control pulse. Very similar curves are found for theory (left panel) and experiment (right panel).

components (compare also Sec. 1.2.8). Excitation when the wavepacket is centered at $FC_{control}$ leads to the greatest decrease in the amplitude of modulation, but not much phase advance or delay. In contrast, excitation when the front/back of the wavepacket is at $FC_{control}$ can result in a phase delay/advance of the wavepacket since the portion remaining on V_1 has its center of gravity behind/ahead with respect to the original undisturbed wavepacket.[24]

This interpretation is verified by our simulations, where we also calculate the change in amplitude and phase induced by the control pulse. Very good agreement is obtained, see Fig. 15. Deviations are due, as before, to orientation averaging included in the experiment but not accounted for in the simulations.

1.4.2. *Ionization to different ionic states*

1.4.2.1. *Preparing electronic wavepackets via SFI*

In this section, we discuss how SFI can launch a "multi-hole" electronic wavepacket, by removing electrons from a superposition of molecular orbitals. A shaped strong-field laser pulse (parametrized in terms of quadratic spectral phase, or chirp) influences the relative contributions of different electronic states in an electronic wave packet consisting of multiple orbital holes and pump-probe measurements. In conjunction with *ab initio* electronic structure calculations, we determine the contribution of three different electronic states in the total electronic and vibrational wavepacket launched via ionization for each pump pulse shape.

We consider the ionization of CH_2I_2 by an intense ultrafast laser pulse. The ionization consists of a mixture of tunnel and multiphoton ionization, with a Keldysh parameter slightly less than 1.[85] The electronic and vibrational wavepackets created via SFI can be probed with a separate "probe" pulse, which transfers the wavepacket between non-dissociative and dissociative potentials as the pump-probe delay is varied. The measurement results are contrasted with similar data for weak-field excitation in the multiphoton regime. The *ab initio* calculations allow us to interpret the fragment ion yield measurements as a function of time delay in terms of vibrational wavepackets on three separate electronic states of the molecular cation.

We perform several IR pump/IR probe and UV pump/IR probe experiments in order to characterize the wavepackets generated via ionization. Figure 16 shows the $CH_2I_2^+$ ion yield as a function of pump-probe delay

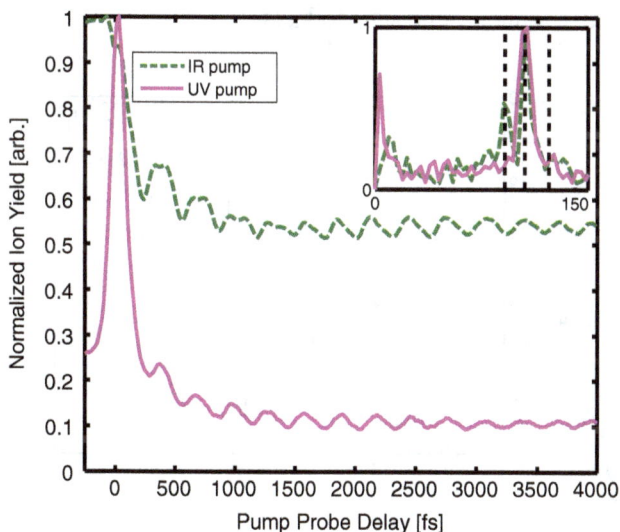

Fig. 16. Comparison of the $CH_2I_2^+$ ion yield vs. pump-probe delay for unshaped UV and IR pump pulses CH_2I_2. The inset shows the normalized value of the Fourier transform of the data vs. frequency in cm^{-1}. The vertical lines are placed at 96, 113 and 130 cm^{-1}. The IR pump data shown here shows peaks at 96 cm^{-1} and 113 cm^{-1}, while data for different IR pulse shapes also shows a peak at 130 cm^{-1} (see Fig. 18).

for both UV and IR pump pulses. Modulations in the CH_2I^+ ion are anti-correlated to the modulations in the $CH_2I_2^+$ ion in both cases, reflecting the probe pulse transferring population between the stable $CH_2I_2^+$ ground state and dissociative states leading to CH_2I^+. The inset shows the Fourier transform of the measured pump-probe signals. The intensities of the IR and UV pump pulses are about 1.3×10^{14} W/cm^2 and 1.6×10^{12} W/cm^2 respectively, with transform-limited pulse durations of 55 fs and 60 fs, respectively. The probe pulses have intensities of about 5×10^{13} W/cm^2 with pulse durations of 60 fs.

A clean oscillation at 112 ± 3 cm^{-1} is found in the UV pump data. It agrees with the calculated I–C–I bending frequency in the ionic ground state of the molecule (108 cm^{-1}). Note that also previous calculations show similar values (e.g., ~ 114 cm^{-1} in Ref. 104). Population transfer to the ionic ground state is expected as the ionization potential is just 0.02 eV below the energy of two UV photons. The modulation in the $CH_2I_2^+$ ion yield as well as in the one of the CH_2I^+ fragment is explained with further population transfer by the probe pulse from the nondissociative ionic ground state to V_4 (see also Fig. 17), as also discussed

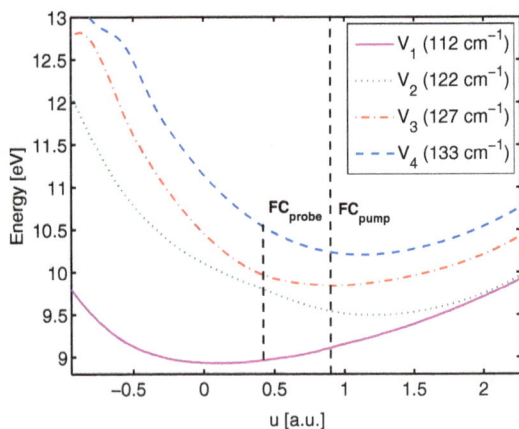

Fig. 17.   This figure shows the four lowest lying ionic potentials of $CH_2I_2^+$. The energy is relative to the neutral ground state minimum. The numbers in brackets give the vibrational frequency near vertical ionization (FC$_{pump}$). The right dotted vertical line indicates the position of the minimum of the neutral ground state (FC$_{pump}$). The left dotted vertical line shows the position of the one-photon resonance between V_1 and V_4(FC$_{probe}$). V_5 and higher PESs are separated by at least another ~ 2.1 eV from V_4 (at $u = 0$) and not shown in the figure.

in earlier publications.[97,98,105,106] V_4 is indirectly dissociative, since it is energetically above the dissociation limit and exhibits nonadiabatic crossings with lower potentials. Thus, any population created on V_4 at FC_{probe} (where V_1 and V_4 come into resonance with the probe) will dissociate slowly via these crossings. Indirect dissociation from V_4 is supported by measurements of fragment yields vs. soft X-ray energy (which show the appearance of CH_2I^+ at 1.1 eV above the ionization potential)[103] and a barrier on V_4 to direct dissociation. The TDM between V_1 and V_4 at FC_{probe} is about 1 a.u., meaning that the potentials can be strongly coupled by the probe pulse.

When using an IR pump pulse instead of the UV pump pulse, the $CH_2I_2^+$ ion yield vs. pump-probe delay signal shows more structure, which is a result of several frequencies beating against one another. The Fourier analysis of the IR pump data yields reveals three important frequencies: 96, 113 and $130\,cm^{-1}$ (all with $\pm 3\,cm^{-1}$). Not all three frequencies are necessarily visible in the pump-probe measurements for a given pump-pulse chirp — e.g., only two are found for the data in Fig. 16. The three frequencies are indicated in Fig. 18, where the spectral content of the derivative of the pump-probe data for a series of IR pump-pulses with different chirp is shown. The advantage of analyzing the derivative and not the original data is that less artifacts from the nonlinear optical response near zero time-delay are obtained but the data is not qualitatively changed. For the curves depicted in Fig. 18, the IR pump-pulse chirp was varied from $-120\,fs^2$ to $+280\,fs^2$ for constant pulse energy. We interpret the presence of these three frequencies in terms of three different vibrational wavepackets launched initially on three separate electronic states.

For an IR pump-pulse, the Keldysh parameter is slightly less than 1, while for the UV pump, it is about 20. Thus, the UV pump-pulse is assumed to be well in the multiphoton regime, whereas the IR pump leads to a mixture of multiphoton and tunnel ionization. The differences in the pump-probe signals of the two pump pulses demonstrate the different character of the ionization process in the two cases. From a comparison of the experimental data with the calculated properties (coordinate-dependent energies, vibrational frequencies and dipole moments) of the different electronic states, we determine which electronic states are excited via

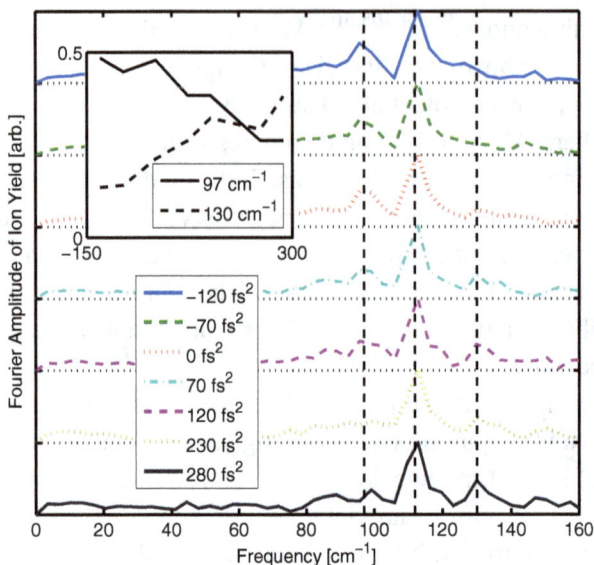

Fig. 18.   Fourier transform of the $CH_2I_2^+$ yield in a series IR pump–IR probe scans for various pump pulse chirps (given in fs^2). Each curve is normalized to the peak at $113\,cm^{-1}$. The frequencies 96, 113 and $130\,cm^{-1}$ are marked by dotted vertical lines. The inset shows the amplitude of the peaks at $96\,cm^{-1}$ and $130\,cm^{-1}$ (both normalized to the $113\,cm^{-1}$ amplitude) vs. chirp (in fs^2).

UV and IR ionization (several states may contribute, i.e., an electronic wavepacket is created, see also Sec. 1.2.9). We also explain how in the case of the IR pump, the pulse chirp affects the content of the electronic wavepacket.

In the IR scans, the modulations at $113\,cm^{-1}$ can be explained in the same way as in the UV pump case: A vibrational wavepacket is created by the pump pulse at FC_{pump} on the ionic ground state, V_1, and further excited to V_4 by the probe pulse when the wavepacket passes through FC_{probe}. Similarly, the observed $130\,cm^{-1}$ oscillations can be interpreted in terms of a vibrational wavepacket evolving on V_4. One comes to this conclusion when looking at the calculated frequency for V_4, which agrees well, but none of the other frequencies for the low-lying potentials (see Fig. 17). As described previously, V_4 is indirectly dissociative, which implies that the $130\,cm^{-1}$ variations in ion yield with pump-probe delay can only be visible if the probe transfers population from V_4 to another, stable PES, (i.e., V_1,

V_2 or V_3 as these are the only states below dissociation limit). Transfer from V_4 to V_1 at FC_{probe} is possible if a wavepacket on V_4 is launched with sufficient displacement from the minimum along u. While it is in principle possible for dynamic Stark shifts to distort V_4 during the pump pulse such that a wavepacket launched at FC_{pump} reaches FC_{probe},[107] we think that it is more likely that V_4 is populated indirectly via V_1. Solving the TDSE for the molecular ion in the tail of the pump pulse shows that a wavepacket initially launched on V_1 can move toward FC_{probe}, allowing population transfer to V_4 at FC_{probe} before the pump pulse is over. A wavepacket launched on V_4 at FC_{probe} by the pump pulse will return to FC_{probe} to be transferred back down to V_1 with the probe pulse every vibrational period. This explains the observed modulations in the $CH_2I_2^+$ ion signal and we therefore assigned the $130\,cm^{-1}$ modulation in our pump-probe measurements to wave packet motion in V_4.

The third frequency of $96\,cm^{-1}$ does not suit any mode for the ground state potential or the bending mode of any molecular cation state near its equilibrium position or close to FC_{pump}. However, the V_1 potential is quite anharmonic as shown by two-dimensional calculations,[108, 109] and the frequency is diminished from $108\,cm^{-1}$ to $96\,cm^{-1}$ for a wavepacket launched on V_1 with a very large displacement (approximately twice the displacement of a wavepacket launched at FC_{pump}). Such a large displacement of a wavepacket can occur if the latter is launched on a higher potential via ionization and relaxes down to V_1 via nonadiabatic couplings. This option is easily possible in the present case as the lower-lying electronic states of the cation come very close to each other in the vicinity of FC_{pump}. Thus, kinetic coupling between the potentials can easily allow for the wavepacket moving from an excited state down to V_1. We suspect that the main contribution to the $96\,cm^{-1}$ modulations come from a wavepacket initially launched on V_2 since a wavepacket initially launched on V_3 would have very little displacement from its minimum (implying slow and low amplitude wavepacket motion) and have to make two nonadiabatic crossings before arriving on V_1.

In the IR pump pulse experiment, we see how an electronic wavepacket is created by a strong field at low frequency. In contrast, the relatively weak field at high frequency of the UV pulse operates far in the multiphoton regime without any tunnel ionization character. Additionally, the photon energy is well above the spacing of the low lying states of the cation. With two UV photons, we can only excite V_1. The energy of V_2 is already too high for a two-photon process. Moreover, the UV pulse is not resonant between V_1 and V_4. Thus, one does not expect a mixture of electronic states to be excited by the UV pump pulse, explaining the lack of $96\,\text{cm}^{-1}$ and $130\,\text{cm}^{-1}$ oscillations in the UV-pump experiment.

Although not all details of the dynamics have been calculated fully (e.g., nonadiabatic coupling between potentials are missing), the observation of three separate vibrational frequencies establishes the excitation of multiple electronic states via ionization, i.e., an electronic wavepacket. In addition to observing the excitation of several electronic states of the cation, we demonstrate that the relative contributions depend on the quadratic spectral phase (or chirp) of the pump pulse and we can thus control the content of the electronic wavepacket. The inset to Fig. 18 shows the variation in the heights of the peaks at $96\,\text{cm}^{-1}$ and $130\,\text{cm}^{-1}$ relative to the $113\,\text{cm}^{-1}$ peak as a function of pump-pulse chirp. Note that the dependence is monotonic with chirp and that the $96\,\text{cm}^{-1}$ and $130\,\text{cm}^{-1}$ peaks have opposite behavior. Concluding, these measurements demonstrate the creation of multiple orbital hole electronic and vibrational wavepackets in a polyatomic molecule created via SFI.

1.4.2.2. *VMI measurements to identify dissociation pathways following SFI*

Following the previous work on bound cationic states of CH_2I_2, we now focus on ionizations to various dissociative electronic states of CH_2BrI^+ that result in the breaking of the C-I bond in the cation, leading to the production of CH_2Br^+.[96] This fragment dominates the time-of-flight mass spectrum and comes from ionic states that correspond to removing an electron from deeper molecular orbitals, rather than just the (highest occupied molecular orbital) (HOMO) or HOMO-1. We measured the velocity distribution of CH_2Br^+ fragments as a function of the shape of the ionizing IR pulse and interpreted the measurements with the help of *ab initio* electronic structure calculations of the parent and fragment cations. The results of these calculations are shown in Fig. 19. As described in detail in Ref. 96, we combined the SA-CASSCF potential energy curves for the I–C–Br bending motion with the more accurate single point MRCI potential

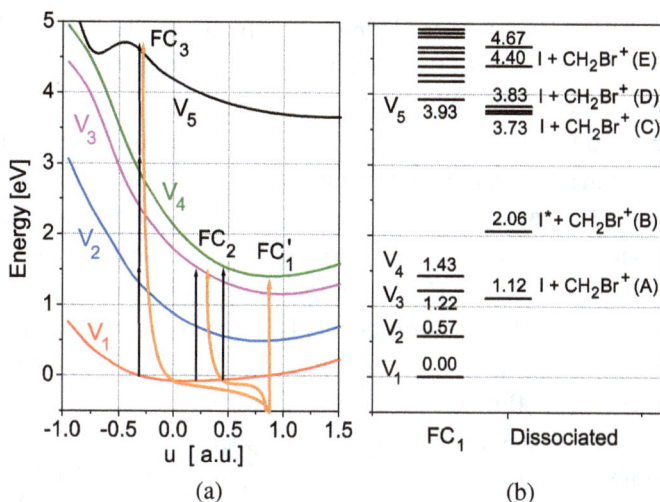

Fig. 19. (color online) Panel a: The energy of the five lowest adiabatic states of CH_2BrI^+ along the bending normal mode coordinate u. The yellow arrows indicate several pathways to excited ionic states via different Franck–Condon points, FC_1 for vertical ionization from the neutral ground state, FC_2 for the close lying V_1/V_3, V_1/V_4 resonances, and FC_3 for the three photon V_1/V_5 resonance. Panel b: The minimum electronic energy of the first few ionic states (left column) and the minimum electronic energy required to create CH_2Br^+ in various electronic states: A, B, C, D and E.

energies for the parent and fragment ions and experimental data (ionization potential of the parent-, and appearance energy of the fragment ion[103]).

The energetics shown in Fig. 19 allows us to identify different possible dissociative ionization pathways and make predictions regarding the kinetic energy release (KER) for each channel. Figure. 19 shows that V_3 (mainly corresponding to the removal of an electron from a lone pair orbital on Br) is the lowest electronic state leading towards dissociation and the production of CH_2Br^+ in channel A. Vertical (direct) ionization to V_3 leads to a maximum (KER) of 100 meV while a vertical (direct) ionization to V_4 has a maximum KER of 310 meV via channel A. However, as Fig. 19 indicates, in addition to direct vertical ionization to V_3 and V_4, ionization to these states can also proceed indirectly via V_1. Ionization to V_1 on the leading edge of the pulse should induce wavepacket evolution on V_1 towards the location FC_2, where V_3 and V_4 are resonant with V_1, and can thus efficiently be populated by the tail of the pump pulse, leading to a maximum KER of 460 meV in channel A. (Let us call this process "indirect" V_3 and V_4

excitation.) While we cannot distinguish experimentally indirect V_3 from indirect V_4 since they have the same total energy, the calculations indicate that indirect V_4 excitation dominates.

Ionizing to V_5 (whose configuration corresponds to a hole in a bonding C-I orbital) can also lead to the production of CH_2Br^+. As with V_3 and V_4, excitation to V_5 can take place either directly (vertical ionization), or indirectly via V_1 at a location, FC_3 (see Fig. 19(a)), close to the inner turning point of the wavepacket in V_1. Direct ionization to V_5 can lead to CH_2Br^+ with four different dissociation channels being energetically accessible: channel A, B, C and D as indicated in Fig. 19(b). The related KERs are 2.81 eV, 1.87 eV, 200 meV and 100 meV, respectively. Molecules excited indirectly to V_5 have a bit more possible dissociation channels resulting in KERs from a few hundred meV (channel E and above) to 3.62 eV (channel A). As will be discussed below, we find that dissociation after ionization to V_5 is dominated by channel A, suggesting that electronic relaxation (curve crossing) takes place prior to dissociation. It should be noted, that fragment ions could also be formed from dissociating neutral states (ion pair states), however, we do not observe any negative ions in any of our measurements.

Figure 20 shows a typical two-dimensional momentum distribution and radial lineouts (with momentum converted to KER) for CH_2Br^+ after

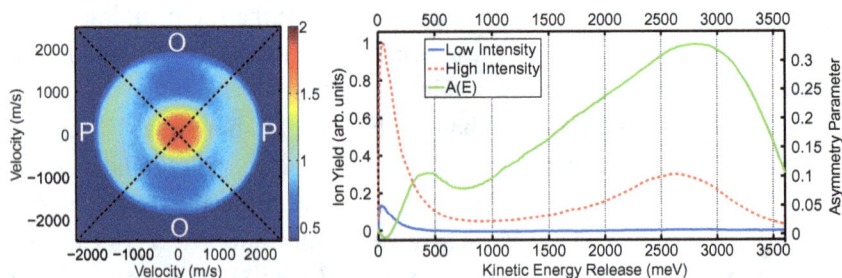

Fig. 20.   (color online) Left panel: 2D velocity distribution of CH_2Br^+ for a single unshaped laser pulse. The x and y axis show the velocity of the detected ions along the x and y direction. The laser polarization was along the x-axis. The total ion yield can be sorted into a slow and fast group of ions, the latter exhibiting a strong tendency to be ejected along the laser polarization. The dashed white lines indicate the borders between the parallel and the orthogonal sectors. Right panel: Radial lineouts for low (blue curve) and high (red curve) intensity, as well as asymmetry parameter (green curve — normalized difference between the yields for fragments parallel to and perpendicular to the laser polarization axis) as a function of KER.

single-pulse excitation. As the creation of each CH_2Br^+ ion from CH_2BrI is essentially a two-body process (neglecting electron recoil and drift in the electromagnetic field), measuring the momentum of either fragment allows us to determine the total KER for this process. However, one has to be careful in comparing the measurements with theory as some energy available following ionization can also be invested in rotational, vibrational and electronic degrees of freedom of the fragments. Furthermore, thermal energy stored in the multiple molecular degrees of freedom prior to ionization and dynamic Stark shifts due to the strong field laser pulse during ionization can lead to high energy tails in the measured KER.[95, 103]

While each ionic state has a different energy, each state can lead to a range of different KERs, given multiple possible dissociation channels. Thus, it is possible for the KERs for different ionic states to overlap, and it is even possible for higher-lying states to have lower KER than lower-lying states. Therefore, it is difficult to assign the different features in the VMI spectrum for a single unshaped laser pulse to specific states without additional information. Hence, we measured the VMI spectrum for different pulse intensities, second-order phases and double pulses in order to establish a clear relationship between features in the VMI spectrum and excitation pathways to several cationic states of the molecule.

In order to single out the KER distributions for ions from excited dissociative states of the cation which are populated indirectly via V_1, we apply the following procedure. We Fourier-transform the CH_2Br^+ KER signal as a function of time delay between an ionizing pump and dissociating probe pulse. There, we consider only the Fourier component corresponding to the vibrational frequency of the wavepacket in V_1. If such a component exists, V_1 is indirectly involved. Here, we treat the Fourier components of the sectors O and P (see definition in Fig. 20) separately. The left panel of Fig. 21 shows the amplitudes and phases for the 95 cm^{-1} Fourier components from the pump-probe data, respectively. The absolute value of the Fourier transform (solid lines) shows two main peaks. This double peak structure becomes even more obvious when looking at the phase associated with each lineout (dashed curves). The phase is generally flat but jumps from 2.1 rad to about 4.1 rad at about 250 meV KER. While the absolute value of the Fourier transform is determined by the amount of population transferred, the phase is determined by the time delay between creating a wavepacket on

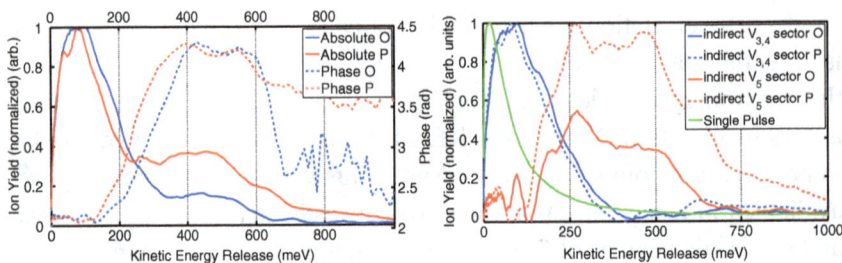

Fig. 21. (color online) Left panel: One-dimensional Fourier transform of velocity map imaged pump-probe signal (CH_2Br^+ yield) evaluated at 95 cm^{-1} as a function of KER. Both sectors (O and P– see Fig. 20) are analyzed separately. The Fourier transformed signal is split into its absolute value (solid lines) and argument (dashed lines). Right panel: Reconstructed KER distributions for the indirect $V_{3,4}$ (blue) and indirect V_5 (red) population, both sectors are shown individually. The green line shows the KER distribution for a single unshaped pulse for comparison.

V_1 (via the pump) and the time of transfer to a dissociative surface (via the probe). Given that there are two peaks in the KER spectrum, and that their phases are different, there must be two separate resonance (FC) positions leading from V_1 to higher-lying (dissociative) states of the cation. Since each peak in the KER distribution exhibits a relatively flat phase (outside of the boundary region where the phase changes rapidly with KER), we reconstruct the two underlying KER distributions by taking the amplitude of the spectrum with the two different phases separately. The results are shown in the right panel of Fig. 21 along with the KER distribution after excitation with the pump pulse only. We are able to assign these two peaks (slow and fast) to indirect V_4 (and V_3) and indirect V_5 respectively, based on numerical integration of time-dependent Schrödinger equation for the molecule in the laser field,[95,96] the cutoff energies from theory, and the asymmetry parameter.

Finally, we consider the effect of second-order spectral phase, or chirp on the direct $V_{3,4}$ versus indirect $V_{3,4}$ contributions in order to test our interpretation of the direct and indirect dissociative ionization pathways. Chirping the pulse increases its duration, allowing for nuclear dynamics on V_1 to play a larger role in the total excited-state population. Numerical and experimental results obtained with chirped pulses are shown in Fig. 22. The agreement between the chirp dependence of the numerical and that of the

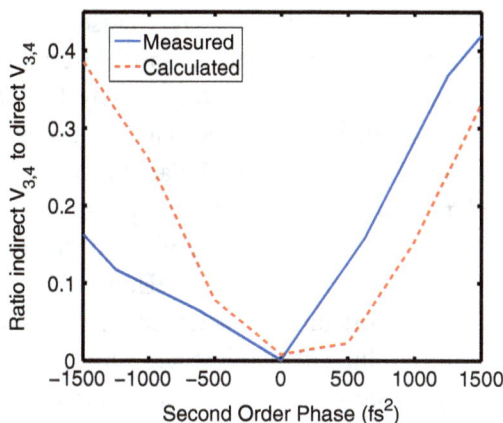

Fig. 22. "Indirect $V_{3,4}$"/"direct $V_{3,4}$" population vs. chirp. The solid graph shows the ratio extracted from our measurements by using the reconstructed direct and indirect $V_{3,4}$ and indirect V_5 distributions (shown in Fig. 21, right panel) as a basis set. The dotted line shows the calculated ratio as a function of chirp.

experimental ratios of "indirect/direct" population transfers confirms the presence of the two excitation pathways.

1.5. Conclusion and Outlook

In this chapter, we have outlined some of the basic ideas underlying laser control of molecular dynamics. We illustrated the concepts with specific examples from a series of halogenated methanes, which lend themselves to interpretation via high-level *ab initio* electronic structure and quantum-dynamics calculations, while offering sufficient complexity to represent a challenge in comparison with atomic or diatomic systems. The fact that these molecules have closely spaced low-lying cationic states makes them well suited for launching electronic wavepackets (superpositions of electronic eigenstates) via SFI, since electrons can tunnel from multiple orbitals whose binding energies are comparable. We discussed how pump-probe spectroscopy in conjunction with detailed electronic structure and quantum-dynamics calculations allows one to interpret the measurements in terms of vibrational wavepackets on multiple electronic states, whose relative contributions could be controlled by the shape of the laser pulse driving the ionization.

A key aspect of SFI with ultrafast laser pulses is the fact that the equilibrium position for many ionic states is shifted from that of the neutral ground state. This means that SFI can impulsively launch a vibrational wavepacket on multiple electronic states. The wavepacket dynamics on these ionic states can be influenced by the application of a second pulse with a controlled intensity, time delay and phase by dressing/mixing cationic states. Specific examples include phase-dependent dissociation, photon locking and spatial hole burning. Understanding all of these control scenarios requires a detailed understanding of molecular electronic structure, spin-orbit coupling, dynamic Stark shifts, dressed states and multiphoton coupling. By combining theory and experiment, it is possible to transfer strong-field control concepts developed for simple systems to more complex ones and advance our ability to control molecular dynamics. In conjunction with parallel developments in molecular alignment, attosecond pulse generation and molecular dynamics modeling, there are many exciting possibilities ahead for directing and observing ultrafast molecular processes.

References

1. D. Strickland and G. Mourou, *Opt. Commun.* **55**, 447 (1985).
2. D. E. Spence, P. N. Kean and W. Sibbett, *Opt. Lett.* **16**, 42 (1991).
3. A. H. Zewail, *J. Phys. Chem. A* **104**, 5660 (2000).
4. M. F. Kling and M. J. Vrakking, *Annu. Rev. Phys. Chem.* **59**, 463 (2008).
5. P. H. Bucksbaum, *Science* **317**, 766 (2007).
6. P. W. Brumer and M. Shapiro, *Principles of the Quantum Control of Molecular Processes.* (Wiley-Interscience, 2003).
7. V. Engel, C. Meier and D. J. Tannor, *Adv. Chem. Phys.* **141**, 29 (2009).
8. K. Bergmann, H. Theuer and B. W. Shore, *Rev. Mod. Phys.* **70**, 1003 (1998).
9. S. A. Rice and M. Zhao, *Optical Control of Molecular Dynamics* (John Wiley & Sons, New York, 2000).
10. M. Dantus, *Ann. Rev. Phys. Chem.* **52**, 639 (2001).
11. M. Shapiro and P. Brumer, *Principles of Quantum Control of Molecular Processes* (Wiley, New York, 2003).
12. I. V. Hertel and W. Radloff, *Rep. Prog. Phys.* **69**, 1897 (2006).
13. P. Nuernberger, G. Vogt, T. Brixner and G. Gerber, *Phys. Chem. Chem. Phys.* **9**, 2470 (2007).
14. G. A. Worth and C. Sanz-Sanz, *Phys. Chem. Chem. Phys.* **12**, 15570 (2010).
15. C. Brif, R. Chakrabarti, and H. Rabitz, *New J. Phys.* **12**, 075008 (2010).
16. B. J. Sussman, *Am. J. Phys.* **79**, 477 (2011).
17. E. T. Sleva, J. I. M. Xavier and A. H. Zewail, *J. Opt. Soc. Am. B* **3**, 483 (1986).
18. D. J. Tannor, R. Kosloff and A. Bartana, *Adv. Chem. Phys.* **101**, 301 (1997).

19. A. Bartana, R. Kosloff and D. J. Tannor, *J. Chem. Phys.* **106**, 1435 (1997).
20. T. Bayer, M. Wollenhaupt, C. Sarpe-Tudoran and T. Baumert, *Phys. Rev. Lett.* **102**, 023004 (2009).
21. D. Tannor, *Introduction to Quantum Mechanics: A Time-Dependent Perspective* (University Science Books, Sausalito, 2006).
22. J. González-Vázquez, L. González, I. R. Sola and J. Santamaria, *J. Chem. Phys.* **131** 104302 (2009).
23. I. R. Sola, B. Y. Chang and H. Rabitz, *J. Chem. Phys.* **119**, 10653 (2003).
24. D. Geißler, P. Marquetand, J. González-Vázquez, L. González, T. Rozgonyi and T. Weinacht, *J. Phys. Chem. A.* **116**, 11434 (2012).
25. L. Fang and G. N. Gibson, *Phys. Rev. Lett.* **100**, 103003 (2008).
26. G. N. Gibson, R. R. Freeman and T. J. McIlrath, *Phys. Rev. Lett.* **67**, 1230 (1991).
27. E. Goll, G. Wunner and A. Saenz, *Phys. Rev. Lett.* **97**, 103003 (2006).
28. H. Goto, H. Katsuki, H. Ibrahim, H. Chiba and K.Ohmori, *Nat. Phys.* **7**, 383 (2011).
29. W. A. Bryan, C. R. Calvert, R. B. King, G. R. A. J. Nemeth, J. D. Alexander, J. B. Greenwood, C. A. Froud, I. C. E. Turcu, E. Springate, W. R. Newell and I. D. Williams, *Phys. Rev. A* **83**, 021406 (2011).
30. H. Niikura, D. M. Villeneuve and P. B. Corkum, *Phys. Rev. Lett.* **92**, 133002 (2004).
31. H. Niikura, D. Villeneuve and P. Corkum, *Phys. Rev. A* **73**, 021402 (2006).
32. B. J. Sussman, M. Y. Ivanov and A. Stolow, *Phys. Rev. A* **71**, 051401 (2005).
33. P. Marquetand, M. Richter, J. González-Vázquez, I. Sola and L. González, *Faraday Discuss* **153**, 261 (2011).
34. J. J. Bajo, J. González-Vázquez, I. Sola, J. Santamaria, M. Richter, P. Marquetand and L. González, *J. Phys. Chem. A* **116**, 2800 (2012).
35. D. Kinzel, P. Marquetand and L. Gonzlez, *J. Phys. Chem. A* **116**, 2743 (2012).
36. M. Richter, P. Marquetand, J. González-Vázquez, I. Sola and L. González, *J. Chem. Theory Computat.* **7**, 1253 (2011).
37. C. C. Marston and G. G. Balint-Kurti, *J. Chem. Phys.* **91**, 3571 (1989).
38. A. Szabo and N. S. Ostlund, *Modern Quantum Chemistry: Introduction to Advanced Electronic Structure Theory* (Dover, Mineola, New York, 1996).
39. B. O. Roos, P. R. Taylor and P. E. M. Siegbahn, *Chem. Phys.* **48**, 157 (1980).
40. B. O. Roos and P. E. M. Siegbahn, *Int. J. Quantum Chem.* **17**, 485 (1980).
41. S. R. Langhoff and E. R. Davidson, *Int. J. Quantum Chem.* **8**, 61 (1974).
42. K. Andersson, P. A. Malmqvist, B. O. Roos, A. J. Sadlej and K. Wolinski, *J. Phys. Chem.* **94**, 5483 (1990).
43. K. Andersson, P.-A. Malmqvist, and B. O. Roos, *J. Chem. Phys.* **96**, 1218 (1992).
44. J. Finley, P.-A. Malmqvist, B. O. Roos and L. Serrano-Andrés, *Chem. Phys. Lett.* **288**, 299 (1998).
45. J. B. Foresman, M. Head-Gordon, J. A. Pople and M. J. Frisch, *J. Phys. Chem.* **96**, 135 (1992).
46. L. González, D. Escudero and L. Serrano-Andrés, *ChemPhysChem.* **13**, 28 (2012).
47. J. F. Stanton and R. J. Bartlett, *J. Chem. Phys.* **98**, 7029 (1993).
48. P. Hohenberg and W. Kohn, *Phys. Rev.* **136**, B864 (1964).
49. W. Koch and M. C. Holthausen, *A Chemist's Guide to Density Functional Theory* (Wiley-VCH, Weinheim, 2001).
50. A. D. Becke, *J. Chem. Phys.* **98**, 5648 (1993).
51. C. Lee, W. Yang and R.G. Parr, *Phys. Rev. B.* **37**, 785 (1988).

52. S. H. Vosko, L. Wilk and M. Nusair, *Can. J. Phys.* **58**, 1200 (1980).
53. E. Runge and E. K. U. Gross, *Phys. Rev. Lett.* **52**, 997 (1984).
54. M. Casida, C. Jamorski, F. Bohr, J. Guan, and D. Salahub, Time-dependent density-functional response theory for molecules, In *Theoretical and Computational Modeling of NLO and Electronic Materials*, eds. S. Karna and A. Yeates (ACS Press, 1996), p. 145.
55. M. Reiher and A. Wolf, *Relativistic Quantum Chemistry* (Wiley-VCH, Weinheim, 2009).
56. C. M. Marian, Spin-Orbit coupling in molecules, in *Reviews in Computational Chemistry*, eds. K. B. Lipicowitz and D. B. Boyd (John Wiley & Sons, 2001).
57. F. Jensen, *Introduction to Computational Chemistry*, 2nd edn. (John Wiley & Sons, 2007), 2 edition.
58. W. Pauli, *Z. Phys. Hadron Nucl.* **43**, 601 (1927).
59. H. A. Bethe and E. E. Salpeter, *Quantum Mechanics of One- and Two-Electron Atoms* (Springer, Berlin, 1957).
60. B. A. He, C. M. Marian, U. Wahlgren and O. Gropen, *Chem. Phys. Lett.* **251**, 365 (1996).
61. B. Schimmelpfennig. *AMFI Atomic Spin-Orbit Mean-field Integral Program* (University of Stockholm, Sweden, 1996).
62. M. Reiher, *Theor. Chem. Acc.* **116**, 241 (2006).
63. M. Douglas and N. M. Kroll, *Ann. Phys.* **82**, 89 (1974).
64. B. A. Hess, *Phys. Rev. A* **33**, 3742 (1986).
65. E. B. Wilson, J. C. Decius and P. C. Cross, *Molecular Vibrations: The Theory of Infrared and Raman Vibrational Spectra* (Dover Publications, 1980).
66. N. Balakrishnan, C. Kalyanaraman and N. Sathyamurthy, *Phys. Rep.* **280**, 79 (1997).
67. M. D. Feit, J. A. Fleck Jr. and A.Steiger, *J. Comput. Phys.* **47**, 412 (1982).
68. M. D. Feit and J. A. Fleck Jr., *J. Chem. Phys.* **78**, 301 (1983).
69. M. D. Feit and J. A. Fleck Jr., *J. Chem. Phys.* **80**, 2578 (1984).
70. J. Stark, *Ann. Phys.* **348**, 965 (1914), ISSN 1521-3889.
71. S. H. Autler and C. H. Townes, *Phys. Rev.* **100**, 703 (1955).
72. C. A. Trallero, *Strong Field Coherent Control* (Stony Brook University, 2007).
73. T. Szakács, B. Amstrup, P. Gross, R. Kosloff, H. Rabitz and A. Lörincz, *Phys. Rev. A* **50**, 2540 (1994).
74. B. Y. Chang, H. Rabitz and I. R. Sola, *Phys. Rev. A* **68**, 031402 (2003).
75. S. Volker, *Annu. Rev. Phys. Chem.* **40**, 499 (1989).
76. T. Brabec and F. Krausz, *Rev. Mod. Phys.* **72**, 545 (2000).
77. V. S. Popov, *Phys.-Usp.* **47**, 855 (2004).
78. L. V. Keldysh, *Sov. Phys. JETP* **20**, 1307 (1965).
79. M. Y. Ivanov, M. Spanner and O. Smirnova, *J. Mod. Opt.* **52**, 165 (2005).
80. S. Augst, D. Strickland, D. D. Meyerhofer, S. L. Chin and J. H. Eberly, *Phys. Rev. Lett.* **63**, 2212 (1989).
81. R. H. Fowler and L. Nordheim, *Proc. R. Soc. Lond. A* **119**, 173 (1928).
82. J. R. Oppenheimer, *Phys. Rev.* **31**, 66 (1928).
83. J. R. Oppenheimer, *Proc. Natl. Acad. Sci. USA* **14**, 363 (1928).
84. A. M. Perelomov, V. S. Popov and M.V. Terent'ev, *Sov. Phys. JETP* **24**, 207 (1967).
85. M. V. Ammosov, N. B. Delone and V. P. Krainov, *Sov. Phys. JETP* **64**, 1191 (1986).

86. F. H. M. Faisal, *J. Phys. B, At. Mol. Phys.* **6**, L89 (1973).

87. H. R. Reiss, *Phys. Rev. A* **22**, 1786 (1980).

88. J. Muth-Böhm, A. Becker, and F. H. M. Faisal, *Phys. Rev. Lett.* **85**, 2280 (2000).

89. A. Becker and F. H. M. Faisal, *J. Phys. B, At. Mol. Opt. Phys.* **38**, R1 (2005).

90. M. Spanner, S. Patchkovskii, C. Zhou, S. Matsika, M. Kotur and T.C. Weinacht, *Phys. Rev. A* **86**, 053406 (2012).

91. M. Kotur, C. Zhou, S. Matsika, S. Patchkovskii, M. Spanner and T. C. Weinacht, *Phys. Rev. Lett.* **109**, 203007 (2012).

92. P. J. Stephens, F. J. Devlin, C. F. Chabalowski and M. J. Frisch, *J. Phys. Chem.* **98**, 11623 (1994).

93. K. A. Peterson, D. Figgen, E. Goll, H. Stoll and M. Dolg, *J. Chem. Phys.* **119**, 11113 (2003).

94. M. J. Frisch, G. W. Trucks, H. B. Schlegel, G. E. Scuseriai, M. A. Robb, J. R. Cheeseman, J. A. Montgomery, T. Vreven, K. N. Kudin, J. C. Burant, J. M. Millam, S. S. Iyengar, J. Tomasi, V. Barone, B. Menuccii, M. Cossi, G. Scalmani, N. Rega, G. A. Petersson, H. Nakatsuji, M. Hada, M. Ehara, K. Toyota, R. Fukuda, J. Hasegawa, M. Ishida, T. Nakajima, Y. Honda, O. Kitao, H. Nakai, M. Klene, X. Li, J. E. Knox, H. P. Hratchian, J. B. Cross, C. Adamo, J. Jaramillo, R. Gomperts, R. E. Stratman, P. Y. Yazyev, A. J. Austin, R. Cammi, C. Pomelli, J. Ochterski, P. Y. Ayala, K. Morokuma, G. A. Voth, P. Salvador, J. J. Dannenberg, V. G. Zakrzewski, S. Dapprich, A. D. Daniels, M. C. Strain, O. Farkas, D. K. Malick, D. Rabuck, K. Raghavachari, J. B. Foresman, J. V. Ortiz, Q. Cui, A. G. Baboul, S. Clifford, J. Cioslowski, B. B. Stefanov, G. Liu, A. Liashenko, P. Piskorz, I. Komaromi, R. L. Martin, D. J. Fox, T. Keith, M. A. Al-Laham, C. Y. Peng, A. Nanayakkara, C. Gonzales, M. Challacombe, P. M. W. Gill, B. G. Johnson, W. Chen, M. W. Wong, C. Gonzales and J. A. Pople, *Gaussian 03 Inc, Pittsburgh PA* (2003).

95. J. González-Vázquez, L. González, S. R. Nichols, T. C. Weinacht and T. Rozgonyi, *Phys. Chem. Chem. Phys.* **12**, 14203 (2010).

96. D. Geißler, T. Rozgonyi, J. González-Vázquez, L. González, P. Marquetand and T. C. Weinacht, *Phys. Rev. A* **84**, 053422 (2011).

97. D. Geißler, T. Rozgonyi, J. González-Vázquez, L. González, S. Nichols and T. Weinacht, *Phys. Rev. A* **82**, 011402 (2010).

98. S. R. Nichols, T. C. Weinacht, T. Rozgonyi and B. J. Pearson, *Phys. Rev. A* **79**, 043407 (2009).

99. P.-O. Widmark, P.-A. Malmqvist and B. O. Roos, *Theor. Chim. Acta* **77**, 291 (1990).

100. B. O. Roos, R. Lindh, P.-Å. Malmqvist, V. Veryazov and P.-O. Widmark, *J. Phys. Chem. A.* **108**, 2851 (2004).

101. G. Karlström, R. Lindh, P.-Å. Malmqvist, B. O. Roos, U. Ryde, V. Veryazov, P.-O. Widmark, M. Cossi, B. Schimmelpfennig, P. Neogrady and L. Seijo, *Comp. Mat. Sci.* **28**, 222 (2003).

102. M. A. Dugan, J. X. Tull and W. S. Warren, *J. Opt. Soc. Am. B.* **14**, 2348 (1997).

103. A. F. Lago, J. P. Kercher, A. Bödi, B. Sztáray, B. Miller, D. Wurzelmann and T. Baer, *J. Phys. Chem. A* **109**, 1802 (2005).

104. X. Zheng and D. L. Phillips, *J. Phys. Chem. A.* **104**, 6880 (2000).

105. D. Geißler, B. J. Pearson and T. Weinacht, *J. Chem. Phys.* **127**, 204305 (2007).

106. B. J. Pearson, S. R. Nichols and T. Weinacht, *J. Chem. Phys.* **127**, 131101 (2007).

107. Z.-H. Loh and S. R. Leone, *J. Chem. Phys.* **128**, 204302 (2008).
108. T. Rozgonyi, J. González-Vázquez, L. González and T. C. Weinacht, Unpublished results.
109. The two chosen coordinates were the Jacobi coordinates in the space spanned by I-C-I bend angle and C-I bond length.

CHAPTER 2

ORIENTATION-SELECTIVE MOLECULAR TUNNELING IONIZATION BY PHASE-CONTROLLED LASER FIELDS

H. Ohmura*

Intense (10^{12}–10^{13} W/cm^2) phase-controlled laser fields consisting of a funda-mental pulse and its second-harmonic pulse induce directionally asymmetric tunneling ionization and resultant orientation-selective molecular ionization in the gas phase. It is demonstrated that orientation-selective molecular ionization induced by phase-controlled $\omega + 2\omega$ laser fields reflect the geometric structure of the highest occupied molecular orbital. This method is robust, being free of constraints such as: the laser wavelength; pulse duration, polarity, and weight of molecules, and thus can be applied to a wide range of molecules. Moreover, this method provides a powerful tool for tracking the quantum dynamics of photoelectrons by using phase-dependent oriented molecules as a phase reference in simultaneous ion–electron detection.

1. Introduction

Since the invention of laser in 1960, the development of laser techniques has progressed rapidly. The tuning ranges of various kinds of laser parameters such as: wavelength, intensity, temporal width, and coher-ence (frequency stability, narrow spectral width) have been increasingly extended or improved. In particular, the advent of techniques to generate intense ultrashort laser pulses enables us to observe nonlinear optical phenomena very easily, leading to the development of various nonlinear laser spectroscopies.

*National Institute of Advanced Industrial Science and Technology (AIST), 1-1-1 Higashi, Tsukuba, Ibaraki 305-8565, Japan. E-mail: hideki-ohmura@aist.go.jp

Multiphoton processes, which are expressed for nonlinear optical phenomena in terms of quantum mechanical perturbation theory, can provide various physical explanations for nonlinear optical effects.[1,2] Recently, a prominent practical application that uses multiphoton processes is the development of multiphoton microscopes.[3] The multiphoton excitation process allows spatial resolutions to be achieved beyond the diffraction limit of conventional optical microscopes and objects to be imaged, that are difficult to observe with conventional microscopes, including the three-dimensional imaging of live tissues. As such, multiphoton microscopes have become powerful tools in the field of bioscience at the state-of-the-art level.[3]

Within conventional perturbation theory in quantum mechanics, as the order of the perturbation increases by one, the number of photons involved also increases by one.[1] Multiphoton processes can ideally describe nonlinear processes at a considerably higher order. However, multiphoton processes involving more than several tens of photons, which are easily induced by intense ultrashort laser pulses, cannot be described adequately by the multiphoton picture. Typical examples of such cases have been found in high-order photoionization[4–18] and high-order harmonic generation (HHG).[19–21] An increase in laser intensity causes a transition from multiphoton ionization (MPI) to tunneling ionization (TI). TI occurs when the laser field suppresses the binding potential of the electron so strongly that the wavefunction of the outermost electron penetrates and escapes the tunneling barrier. Keldysh theory successfully described the transition between MPI and TI in 1963, soon after the invention of laser.[4] This theory explains that the multiphoton picture cannot provide insight into the physical properties of nonlinear effects involving a large number of photons.[4] Recent studies have revealed that TI occurs mainly in the attosecond (as) time region (1 as $= 10^{-18}$ s), when the electric field of the laser reaches its maximum value owing to a highly nonlinear optical response.[19–21] Analogous to scanning tunneling microscopy (STM), where quantum-tunneling phenomena are applied to observe atomic-scale objects in the space domain, the use of TI induced by intense laser fields enables the observation of ultrafast changes in the time domain. Therefore, TI is one of the most important and fundamental phenomena for measuring and controlling physics in the attosecond time regime.[19–21]

HHG, which converts many absorbed photons into a single photon with an energy equivalent to the sum of the absorbed photons, exhibits remarkable behavior that cannot be described by perturbation theory.[19-21] In other words, HHG spectra do not obey the behavior expected by conventional perturbation theory. Corkum's recollision model has successfully described the observed behaviors.[19-21] As a consequence of accelerated motion synchronized with an oscillating laser field, electrons induced by TI are pulled away from, pulled back near to, and recollided with parent ions within one optical cycle. This recollision process plays an important role, not only in the coherent generation of soft X-rays, but also of attosecond light pulses.[19-21] Corkum's recollision model emphasizes the importance of coherent motion of electrons synchronized with an oscillating laser field, which is not accounted for in the conventional multiphoton model.

Such coherent motions of electrons synchronized with an oscillating laser field are strongly affected by the laser's phase. Therefore, so-called coherent or quantum control, which is the direct manipulation of the wavefunction and its quantum dynamics through the coherent nature of a laser field, is expected to be a powerful tool to control the coherent motion of electrons (for reviews see Refs. 22 and 23). Toward this goal, we have investigated the coherent control of molecular TI processes using phase-controlled, two-color laser pulses consisting of a fundamental pulse and its second-harmonic pulse (the $\omega + 2\omega$ pulse).[24-64]

In this review, we report the use of phase-controlled laser fields to achieve quantum control of molecular TI in the space domain and the resultant orientation-selective molecular ionization in the gas phase.[57-64] First, the basic properties of TI induced by intense laser fields for atoms and molecules are described in Sec. 2. Then, the characteristics of $\omega + 2\omega$ laser fields and directionally asymmetric TI induced by $\omega + 2\omega$ laser fields, as well as the principles of orientation-selective molecular ionization, are described in Sec. 3. After that, explanations of the experimental apparatus, method for generation of $\omega + 2\omega$ laser fields, and detection of oriented molecules are presented in Sec. 4. Experimental results and discussions of directionally asymmetric molecular TI induced by the $\omega + 2\omega$ laser field are discussed in Sec. 5. Finally, a brief summary is provided.

2. Photoionization Induced by Intense Laser Fields

2.1. *MPI in standard perturbation theory*

N-photon processes can be described by taking nth-order perturbation into consideration in quantum theory, where n is an integer. In this section, we consider MPI, where the energy of multiphoton absorption exceeds the ionization potential energy, E_{IP} and an electron is emitted (Fig. 1(a)). According to Fermi's golden rule, in the standard perturbation expansion of the time-dependent Schrödinger equation, the MPI rate W_{MPI} is given by,[1,2]

$$W_{MPI} = \frac{2\pi}{\hbar} \left(\frac{2e^2}{\varepsilon_0 c} \right)^n I^n \left| \sum_{i,j,\dots,k} \frac{\langle E|r|k\rangle \langle k|r|j\rangle \cdots \langle i|r|g\rangle}{(E_k - E_g - (n-1)\hbar\omega) \cdots (E_i - E_g - \hbar\omega)} \right|$$

$$\times \delta(E - E_g - n\hbar\omega) \tag{1}$$

where $\hbar$ is Planck's constant divided by 2π, e is the elementary charge, ε_0 is the vacuum permittivity, c is the speed of light, I is the laser intensity,

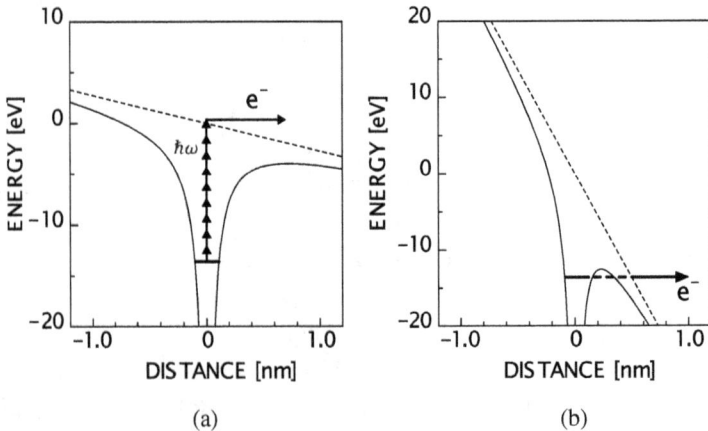

Fig. 1. (Solid curves) Coulomb potential of a hydrogen atom ($E_{IP} = 13.6\,\text{eV}$) in the electric field of a laser at two different laser intensities I. (a) $I = 1.0 \times 10^{12}\,\text{W/cm}^2$ ($\gamma = 3.37$). Ladder of vertical arrows represents multiphoton ionization consisting of nine photons ($\hbar\omega = 1.55\,\text{eV}$, $\lambda = 800\,\text{nm}$). (b) $I = 1.0 \times 10^{14}\,\text{W/cm}^2$ ($\gamma = 1.07$). Horizontal arrow in (b) shows tunneling ionization by escaping through the potential barrier, which is suppressed by the intense laser field. Dashed lines show the potential for electric fields of the laser only. Horizontal thick solid lines show the energy level of the ground state.

ω is the laser frequency, n is the minimum number of photons that exceed the ionization threshold, $|g\rangle$ is the ground state with associated energy E_g, and $|E\rangle$ is the continuum state with associated energy E (unless indicated, we use MKS units throughout the paper). E_g corresponds to the ionization potential E_{IP}. W_{MPI} is proportional to the nth power of the laser intensity I. Energy conservation leads to photoelectron energy given by $E = n\hbar\omega - E_{IP}$. When MPI includes a resonant optical transition to an intermediate state, atoms and molecules are effectively ionized by relatively low laser intensity. This phenomenon is called resonance-enhanced MPI (REMPI). REMPI is widely used not only in the field of atomic and molecular physics but also in the field of microchemical analysis and analytical chemistry.

At intensities of 10^{12}–10^{13} W/cm^2, electrons do not cease to absorb photons, even after $n\hbar\omega$ exceeds the ionization potential. For such conditions, the corresponding photoelectron energy is modified as $E = (n + s)\hbar\omega - E_{IP}$, where s is the number of additionally absorbed photons. This is called above-threshold ionization (ATI).[65,66] A photoelectron spectrum resulting from ATI shows a series of discrete peaks separated by the photon energy $\hbar\omega$ (see Fig. 17(a) in Sec. 5.1.2). MPI can provide a physical picture where atoms and molecules are ionized by absorbing several photons simultaneously in a high-density photon field. Although multiphoton processes can be ideally applied for the interpretation of high-order nonlinear optical phenomena, MPI does not provide a sufficient physical description of multiphoton processes involving more than several tens of photons.

2.2. Keldysh theory: From MPI to TI

An insightful theory with respect to photoionization induced by intense laser fields has been suggested by Keldysh.[4] By using first-order perturbation theory, he treated the photoionization rate between the bounded ground state of a hydrogen atom and the Volkov continuum state, which includes oscillatory motion of free ionizing electrons induced by a linearly polarized electric field, as final states instead of simple continuum states,

$$\psi_p = \exp\left\{\frac{i}{\hbar}\left[\tilde{p}(t) \cdot r - \int_0^t \frac{1}{2m}\tilde{p}(\tau)^2 d\tau\right]\right\}, \quad \tilde{p}(t) = p + \frac{eF}{\omega}\sin\omega t,$$

(2)

where F is the electric field amplitude, m is the mass of the electron, and p is the momentum of the electron. In spite of the first-order perturbation expansion, the ionization rate W_{Keldysh} includes the form of the high-order nth multiphoton absorption[4]

$$W_{\text{Keldysh}} = \frac{2\pi}{\hbar} \int \frac{d^3 p}{(2\pi\hbar)^3} |L(p)|^2 \sum_{n=-\infty}^{\infty}$$

$$\times \delta\left(\tilde{E} + \frac{p^2}{2m} - n\hbar\omega\right), \quad \tilde{E} = E_{\text{IP}} + U_p,$$

$$L(p) = \frac{1}{2\pi} \oint V_0\left(p + \frac{eF}{\omega}u\right)$$

$$\times \exp\left\{\frac{i}{\hbar\omega} \int_0^u \left[E_{\text{IP}} + \frac{1}{2m}\left(p + \frac{eF}{\omega}v\right)^2\right] \frac{dv}{(1-v^2)^{1/2}}\right\} du,$$

$$(3)$$

where $U_p = e^2 F^2/4m\omega^2$ is the ponderomotive energy. Moreover, when the Keldysh adiabatic parameter $\gamma = \sqrt{E_{\text{IP}}/2U_p}$ is $\ll 1$, which corresponds to the case of low frequency and very strong laser fields (adiabatic approximation), the ionization rate can be transformed to

$$W_{\text{Keldysh},\gamma \ll 1} = \frac{\sqrt{6\pi}}{4} \frac{E_{\text{IP}}}{\hbar} \left(\frac{eF\hbar}{m^{1/2} E_{\text{IP}}^{3/2}}\right)^{1/2}$$

$$\times \exp\left\{-\frac{4}{3} \frac{\sqrt{2m} E_{\text{IP}}^{3/2}}{e\hbar F}\left(1 - \frac{m\omega^2 E_{\text{IP}}}{5e^2 F^2}\right)\right\}. \quad (4)$$

As $\omega \to 0$, Keldysh theory found that Eq. (4) coincides with the well-known formula of TI for a hydrogen atom in a static constant electric field[67-69]:

$$W_{\text{static TI}} = 4\omega_{au}\left(\frac{F_{au}}{F}\right) \exp\left\{-\frac{2}{3}\left(\frac{F_{au}}{F}\right)\right\},$$

$$\omega_{au} = \frac{1}{(4\pi\varepsilon_0)^2} \frac{me^4}{\hbar^3} \quad \text{and} \quad F_{au} = \frac{1}{(4\pi\varepsilon_0)^3} \frac{m^2 e^5}{\hbar^4}, \quad (5)$$

where ω_{au} and F_{au} are the atomic units of the frequency and the electric field, respectively. As mentioned in Sec. 1, TI occurs when the electric field suppresses the Coulomb potential so strongly that the wavefunction of the outermost electron penetrates and escapes the tunneling (Fig. 1(b)). Keldysh theory has been found to provide continuous connection between MPI and TI, and the Keldysh adiabatic parameter γ can be used to judge whether a given observed phenomenon is of the MPI or TI type. If $\gamma > 1$, MPI is dominant; if $\gamma < 1$, then TI is dominant.[4] These facts indicate that the Volkov continuum state accounts for the main effect of the electric field — the acceleration of the free electron — and becomes a more adequate basis set for understanding the physical picture, including both MPI and TI. The adiabatic approximation $\gamma < 1$ corresponds to the physical situation where the wavefunction of the outermost electron's penetration through the tunneling barrier can follow the temporal change in the electric field. Since the adiabatic approximation in Keldysh theory was found to lead to the formula of TI in a constant electric field, the TI rate in intense laser fields is easily calculated by substituting $F \rightarrow F\cos(\omega t)$ in Eq. (5). Perelomov, Popov, and Terent'ev developed an expression for the TI rate for arbitrary atoms, as well as arbitrary γ, on the basis of a Green's function method; this theory is known as PPT theory.[5] As a simple extension of PPT theory, Ammosov, Delone, and Krainov derived an expression of TI rate for arbitrary states in arbitrary atoms and atomic ions[6]:

$$
W_{ADk} = \left(\frac{3}{\pi^3}\right)^{1/2} \frac{(2l+1)(l+|m|)!}{(|m|)!(l-|m|)!} \times \left(\frac{e}{(n^{*2}-l^{*2})^{1/2}}\right)^{|m|+3/2}
$$

$$
\times \left(\frac{n^*+l^*}{n^*-l^*}\right)^{l^*+1/2} \times \frac{Z^2}{n^{*3}} \left(\frac{4eZ^3}{Fn^{*3}(n^{*2}-l^{*2})^{1/2}}\right)^{2n^*-|m|-3/2}
$$

$$
\times \exp\left(-\frac{2Z^3}{3n^{*4}F}\right) \tag{6}
$$

where n^* is the effective principal quantum number of the state, $l(l^*)$ and m are the (effective) orbital quantum number of the state and its projection, and Z is the charge of the atomic residue. This theory is well-known as ADK theory and is commonly used to compare between the theory and experiment on absolute values and F-dependence for ionization rate.

Keldysh theory has been studied by comparing it with standard perturbation theory and nonperturbative theory of electron scattering in intense laser fields.[70,71] Keldysh's treatment is perturbation expansion not with respect to the applied laser field but to the binding potential $V(r)$ for the electron, and the Keldysh approximation can be regarded as an ansatz rather than a leading term in the perturbation series.[71,72] Keldysh theory has been a benchmark in the theory of strong-field physics.

Keldysh theory has been further developed by researchers to compare its output with experimental results from real atoms.[73,74] One attribute of Keldysh theory is that the nth multiphoton process can be included even in first-order perturbation. The accuracy of each n-photon transition rate increases order-by-order in the high-order correction. Faisal and Reiss have established a rigorous basis for an extended version of Keldysh theory in which systematic high-order corrections can be applied to the Keldysh term[72,73]; this basis is well known as Keldysh–Fisal–Ress (KFR) theory.

2.3. *Characteristics of TI*

In this section, characteristics of TI in intense laser fields are introduced under the adiabatic approximation ($\gamma < 1$) by applying Eq. (5) to the hydrogen atom ($E_{IP} = 13.6\,\text{eV}$) with a laser wavelength λ of 800 nm and an intensity I of $1.0 \times 10^{14}\,\text{W/cm}^2$.

Figure 2 depicts the tunneling ionization rate as a function of laser intensity I by using $F = \sqrt{2I/\varepsilon_0 c}$. The TI rate drastically increases as the

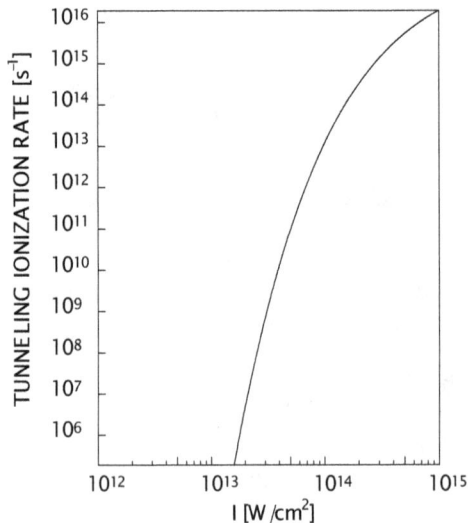

Fig. 2.   Tunneling ionization rate as a function of the laser intensity I for a hydrogen atom as calculated by Eq. (5).

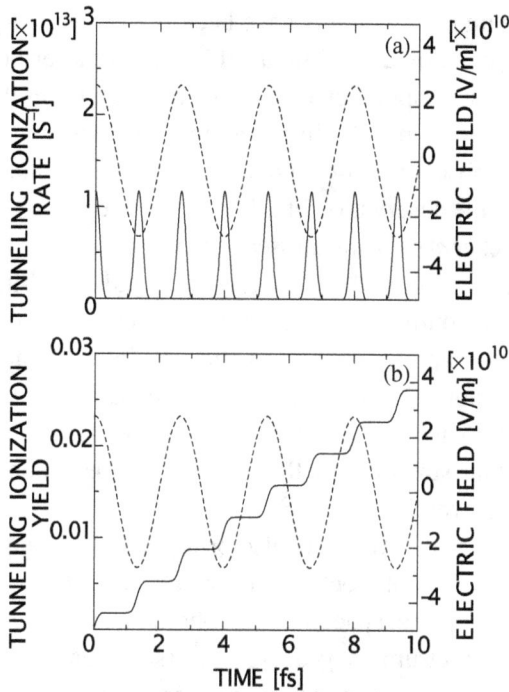

Fig. 3. Temporal behaviors of tunneling ionization for a hydrogen atom ($E_{\rm IP} = 13.6\,{\rm eV}$) in the linearly polarized electric field of a laser. (a) Tunneling ionization rate as a function of time; (b) Tunneling ionization yield as a function of time. A quasistatic TI model was used with the following calculation parameters: $\lambda = 800\,{\rm nm}$, $I = 1.0 \times 10^{14}\,{\rm W/cm}^2$, $\gamma = 1.07$. Dashed curves represent the electric field.

laser intensity increases because of its exponential dependence. In the adiabatic approximation ($\gamma < 1$), the temporal behavior of the TI rate in intense laser fields is easily calculated by substituting $F \rightarrow F \cos(\omega t)$ in Eq. (5).

Figure 3(a) shows the TI rate as a function of time. TI occurs mainly in the attosecond time region, when the electric field reaches its maximum values owing to highly nonlinear optical response, and the ionization rate profile consists of sharp peaks. As a result, the ionization yield, which is calculated by integrating the ionization rate over time, increases stepwise as a function of time (Fig. 3(b)). TI behavior contrasts with that of MPI: the MPI rate per optical cycle is much smaller than that of TI, and ionization proceeds slowly over many optical cycles so that the ionization yield can be regarded as increasing linearly with respect to time. Analogous to STM, where quantum tunneling effects have been used to magnify atomic-scale objects in the space domain, TI induced by intense laser fields enables us to observe ultrafast temporal changes that are much faster than one optical cycle, in the time domain.[19-21] In real atomic and molecular systems, it has been difficult to observe attosecond TI bursts experimentally. Recently, a successful real-time observation of a TI process (step-like TI behavior) was made by Uiberacker *et al.* using attosecond extreme ultraviolet (XUV) pulse pumping and a near infrared (NIR), few-cycle pulse probing technique.[74] In addition, some controversy exists regarding the range of γ values over which the adiabatic approximation is valid. The intermediate regime $\gamma \sim 1$ is referred to as nonadiabatic TI,[9] but nonadiabatic TI in helium with γ ranging from 1.45 to 1.17 have been experimentally explored with an attosecond angular streaking method, and it has been concluded that the TI model is valid even for $\gamma > 1$.[75]

The transition from MPI to TI also brings about a change in the photoelectron spectrum. The photoelectron spectrum of MPI, which consists of a series of discrete peaks spaced by the photon energy $\hbar\omega$ (ATI), changes to a continuum spectrum as γ decreases[8] (see Figs. 17(a) and 17(b) in Sec. 5.1.2). The photoelectron spectrum generated by TI can be qualitatively reproduced by a quasistatic TI model.[7] The quasistatic TI model consists of two steps. In the first step, TI gives rise to photoelectrons with the assumption of zero velocity $v(t_0) = 0$ at ionization time t_0, and with a tunneling rate $W(t)$ that depends on the instantaneous electric field given by Eq. (5). In the second step, the center-of-mass motion of the electron wave packet is driven by oscillating laser fields. The approximation is that after ionization, the Coulomb attraction between the photoelectron and the parent ion is so small compared to the laser fields that Coulomb attraction can be neglected. By solving the Newton equation of motion for the photoelectron driven by the single-frequency, linearly polarized laser field,

$$m\frac{dv(t)}{dt} = -eF\cos(\omega t) \tag{7}$$

and using an initial velocity $v(t_0) = 0$, we obtain the following solution for velocity $v(t)$:

$$v(t) = -\frac{eF}{m\omega}[\sin(\omega t) - \sin(\omega t_0)]. \tag{8}$$

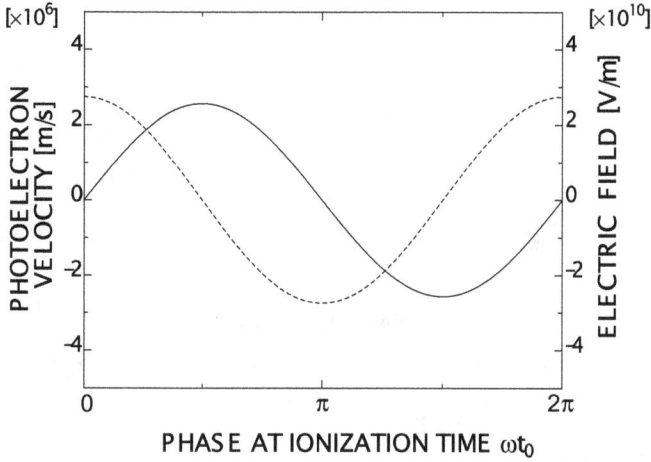

Fig. 4. Velocity of photoelectrons generated by tunneling ionization as a function of phase at ionization time. A quasistatic TI model employing the same calculation parameters as those listed in the caption of Fig. 3 are used. The dashed curve represents the electric field amplitude of the laser.

The first term represents the oscillatory motion synchronized with the laser field. After the laser pulse passes the interaction region, the first oscillatory term driven by the laser field in Eq. (8) disappears, and the final electron velocity $v_{final}(t_0)$ can be expressed by the following equation:

$$v_{final}(t_0) = \frac{eF}{m\omega} \sin(\omega t_0). \tag{9}$$

The relationship between the electric field amplitude and final electron velocity as a function of phase at ionization time t_0 is depicted in Fig. 4. The electron velocity generated by TI is dependent on the ionization time t_0. The majority of photoelectrons generated by the maxima of the electric field at $\omega t_0 = 0, \pi$ have a velocity of zero. The highest-velocity photoelectron is generated at $\omega t_0 = \pi/2, 3\pi/2$. The kinetic energy of each photoelectron is

$$E_k = \frac{1}{2}m v_{final}^2 = 2U_p \sin^2(\omega t_0). \tag{10}$$

The maximum photoelectron kinetic energy is limited to $2U_p$. By summing the TI rate at ionization times that give rise to equal kinetic energy, we can obtain the photoelectron spectrum generated by TI as a function of kinetic energy (Fig. 5). This quasistatic TI model can qualitatively

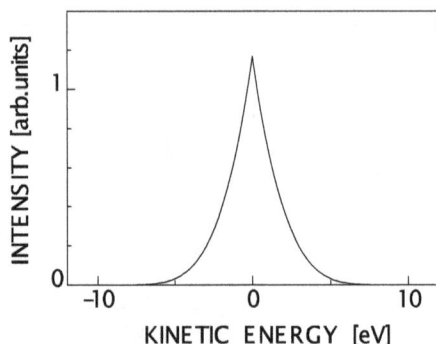

Fig. 5. Photoelectron spectrum for tunneling ionization as a function of kinetic energy. A quasistatic TI model employing the same calculation parameters as those listed in the caption for Figure 3 was used. Positive (negative) kinetic energy indicates a positive (negative) photoelectron velocity.

reproduce the exponentially decreasing continuum spectrum characteristic of photoelectrons generated by TI; this spectrum is commonly observed in experiments.[7] The validity of the assumption $v(t_0) = 0$ has been confirmed by full quantum simulations.[10] As discussed in Sec. 5.1.2, however, although the quasistatic TI model is valid for high-energy photoelectrons, a Coulomb force correction is required for low-energy photoelectrons.[45,46,63]

2.4. Molecular TI

When considering molecular TI, the outermost electronic cloud for molecules is not spherically symmetric compared to that for atoms, because molecules consist of more than one atom. This asymmetry leads to an angular dependence of the TI rate between the molecular axis and the polarization of laser fields. The ADK model, which is commonly used as the TI theory for atoms, has been extended by Tong and Lin to treat molecular systems.[11,12] The ADK model for atoms is derived from the wavefunction of a valence electron that initially has well defined spherical harmonics $Y_{lm}(\mathbf{r})$. For atoms, the wavefunction of the valence electron at large distances away from the potential barrier, where tunneling occurs, can be written as $\Psi^m(r) = C_l F_l(r) Y_{lm}(\mathbf{r})$, where $F_l(r \rightarrow \infty) \approx r^{Z_c/\kappa - 1} e^{-\kappa r}$ with Z_c as the effective Coulomb charge, $\kappa = \sqrt{2E_{IP}}$. The valence orbital for molecules is referred to as the highest occupied molecular orbital (HOMO). The HOMO of molecules is commonly expressed as the linear combination of atomic orbitals on the basis of multi-center expression. However, multi-center expression is not directly applied in the ADK model.[11,12] To employ the ADK formula directly for molecules, expansion of the molecular wavefunctions for the asymptotic region in terms of summation of spherical harmonics in a one-center expression is required. For molecules, the HOMO in the tunneling region can be written as $\Psi^m(r) = \sum_l C_l F_l(r) Y_{\text{lm}}(\mathbf{r})$. The coefficients C_l are obtained by fitting the asymptotic molecular wavefunction. Once the coefficients C_l are available, the TI rate for molecules can be calculated on the basis of the ADK model.

Figure 6 depicts the molecular structures and isocontours of the HOMO determined by *ab initio* calculations using the Gaussian 03W software

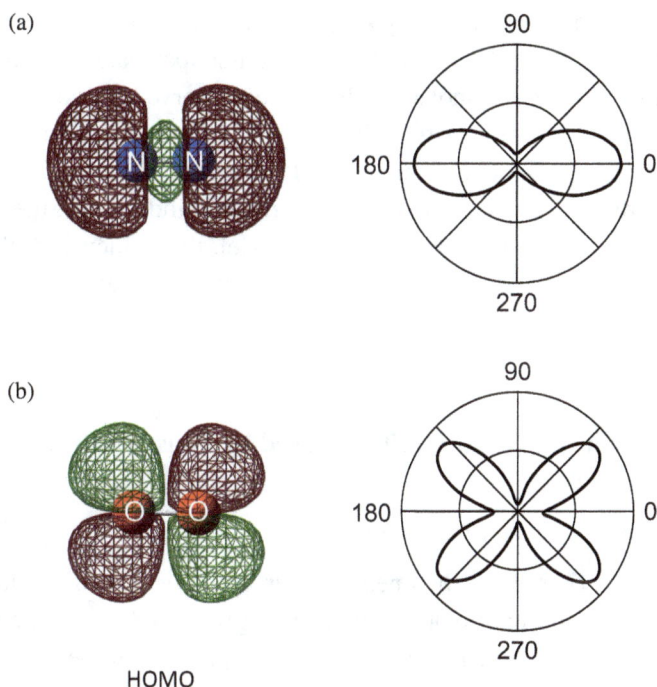

Fig. 6. (Left panel) Molecular structures and isocontours of HOMO of (a) the nitrogen molecule and (b) the oxygen molecule. The shading indicates the sign of the wavefunction. (Right panel) Polar plots of tunneling ionization rate as a function of angle between laser polarization and molecular axis for (a) the nitrogen molecule and (b) the oxygen molecule calculated by the molecular ADK model. The calculated data are taken from Ref. 12.

package[76] and polar plots of tunneling ionization rate as a function of angle between laser polarization and molecular axis for (a) the nitrogen molecule and (b) the oxygen molecule calculated by the molecular ADK model.[12] The HOMO of the N_2 molecule has σ_g symmetry and that of the O_2 molecule has π_g symmetry. An immediate consequence of molecular ADK theory at first glance is that the TI rate reflects the geometric structure of the HOMO. From this fact we can interpret that photoelectrons are much more strongly extracted via tunneling from the large-amplitude lobe of the HOMO along the opposite direction of the electric field vector. As a consequence of the angular dependence of the TI rate, molecules aligned in a certain direction are selectively ionized in a randomly oriented gas-phase molecular ensemble, and when photofragmentation is induced by TI, the photofragment-emission pattern reflects the structure of the molecular orbital. For example, a butterfly-shaped pattern reflecting the structure of the π orbital in O_2 molecules and a dumbbell-shaped pattern reflecting the σ orbital structure in N_2 molecules have been observed in two-dimensional photofragment-emission pattern imaging.[13,14] Recently, deviations from the observed angular dependence of the TI rate from molecular ADK theory have been discussed quantitatively to confirm the validity of the theory.[15−18] To further improve the molecular ADK model, the influence of the linear Stark effect has been discussed and tested (Stark-corrected molecular ADK theory).[49,77−80]

3. Directionally Asymmetric TI Induced by Phase-controlled Laser Fields

3.1. *Phase-controlled laser fields*

The total electric field of the linearly polarized optical fields of the fundamental and second-harmonic frequencies is given by $F(t) = F_1 \cos(\omega t) + F_2 \cos(2\omega t + \phi)$, where F_1 and F_2 are the amplitudes of the electric fields and ϕ is the relative phase difference between the fundamental and the second-harmonic light. The amplitude of the electric field in the positive (negative) direction is about twice that in the negative (positive) direction when $\phi = 0(\pi)$ (Figs. 7(a) and 7(b)). The phase-controlled $\omega + 2\omega$ laser fields have a characteristic, phase-dependent asymmetric waveform, in

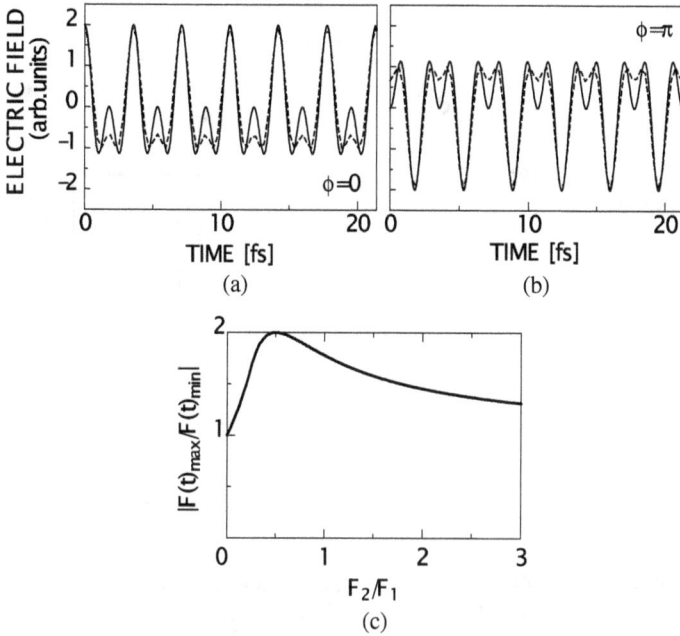

Fig. 7.   (a) Waveforms of phase-controlled, two-color $\omega + 2\omega$ laser fields at a relative phase difference of (a) $\phi = 0$ and (b) $\phi = \pi$ (solid line, $F_2/F_1 = 1.0$; dotted line, $F_2/F_1 = 0.5$). (c) $|F(t)_{max}/F(t)_{min}|$ at $\phi = 0$ as a function of F_2/F_1.

contrast to single-frequency laser fields, which have symmetric waveforms. Figure 7(c) shows the degree of asymmetry $|F(t)_{max}/F(t)_{min}|$ plotted as a function of the ratio F_2/F_1 at $\phi = 0$. $|F(t)_{max}/F(t)_{min}|$ reaches a maximum at $F_2/F_1 = 0.5$.

Phase-controlled $\omega + 2\omega$ laser fields have been investigated as a means for coherent control or quantum control, which is the direct manipulation of the wavefunction and its quantum dynamics through the coherent nature of a laser field (for reviews see Refs. 22 and 23). For weak laser fields ($<10^{12}$ W/cm^2), a phototransition scheme can be used to explain the phase-dependent phenomena induced by phase-controlled $\omega + 2\omega$ laser fields. Quantum interference occurs between a one-photon transition of the second-harmonic photon and a two-photon transition of the fundamental photon (Fig. 8(a)). In this scheme, because the final states differ in parity owing to the different selection rules for the one- and two-photon transitions, the interference between the two transitions

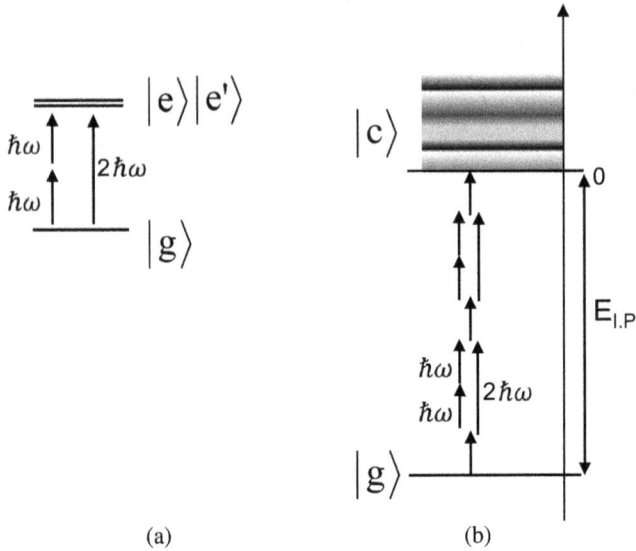

Fig. 8. (a) Schematic of the energy diagram corresponding to the interference between a one-photon transition induced by second-harmonic light and a two-photon transition induced by fundamental light. $|g\rangle$ is the ground state, $|e\rangle$, $|e'\rangle$ are the excited states, and $|c\rangle$ is the continuum state. In the case of MPI (b), the two-path interference can occur multiple times (twice in the figure), leading to multi-pathway interference.

induces the breaking of spatial symmetry.[22,23] This type of interference has been studied elsewhere, including angular distribution of photoelectrons in atoms[24–27] and molecules,[28] and in photocurrents in semiconductors.[29–31] For extending the application of the $\omega + 2\omega$ scheme, the implication of external electric fields that mix states of opposite parity,[32–35] as well as the photodissociation of molecules,[36–38] have been studied. In the case of MPI, two-pathway interference can involve several times at any intermediate steps in MPI (Fig. 8(b), twice are included) so that multi-pathway interference is possible. However, the assignment of a transition pathway is increasingly complicated, so it is hard for us to grasp the physical picture of multi-pathway interference at weak laser fields.

In contrast, for intense laser fields ($> 10^{12}$ W/cm^2), a scheme involving electric fields that induce motion of charges or dipoles has been presented. The asymmetric electric fields directly induce the motion of electrons or dipoles with asymmetric directionality. The manipulation of the directional asymmetry induced by intense phase-controlled $\omega + 2\omega$ laser fields has been investigated for photoelectrons in atoms[39−43] and for photoelectrons and photodissociation in molecules.[44−49] The application of the $\omega + 2\omega$ scheme to molecules leads to effective control of molecular orientation with discrimination of the molecules' head–tail order; such control is impossible to achieve with a monochromatic laser field with a symmetric waveform.[50−56]

3.2. *Directionally asymmetric TI (atoms)*

In this section, we describe TI induced by phase-controlled $\omega + 2\omega$ laser fields on the basis of the same quasistatic TI model[41,42,45,46] used to describe single-frequency laser fields, and we compare single-frequency laser fields with phase-controlled double-frequency laser fields. To modify the quasistatic TI model for phase-controlled $\omega+2\omega$ laser fields, we perform the simple substitution of $F \cos(\omega t) \rightarrow F_1[\cos(\omega t) + r \cos(2\omega t + \phi)]$ as described in Sec. 2.3, where r is the ratio F_2/F_1. The hydrogen atom ($E_{IP} = 13.6\,\text{eV}$) and a laser with wavelength of $400 + 800\,\text{nm}$ and a total intensity $I = I_1 + I_2 = 1.0 \times 10^{14}\,\text{W/cm}^2$ ($I_1 = 5.0 \times 10^{13}\,\text{W/cm}^2$, $I_1 = 5.0 \times 10^{13}\,\text{W/cm}^2$) are considered in the calculation. Figure 9 depicts the temporal behavior of TI in the case of a linearly polarized, phase-controlled $\omega + 2\omega$ laser field at $\phi = 0$ calculated by Eq. (5). Reflecting that the amplitude of the electric field in the positive direction is about twice that in the negative direction, TI occurs at the time when the electric field reaches its peak in the positive direction by high-order nonlinear optical response. The directionality is flipped by changing ϕ from 0 to π. Phase-controlled $\omega + 2\omega$ laser fields can induce the directionally asymmetric TI, which have asymmetric waveforms, in contrast to the symmetric waveforms observed for single-frequency laser fields. The resultant period of stepwise ionization yield for the phase-controlled $\omega + 2\omega$ laser fields increases two times as compared to the case of single-frequency laser fields (Fig. 9(b)). Intense phase-controlled $\omega + 2\omega$ laser fields enable us to manipulate the directionality of TI in the attosecond time region. Therefore, we can say that directionally asymmetric TI induced by phase-controlled $\omega + 2\omega$ laser fields is the manipulation of the electron in the spatiotemporal domain.

The most intriguing effect induced by the asymmetric waveforms of the phase-controlled $\omega + 2\omega$ laser fields can be seen in their photoelectron

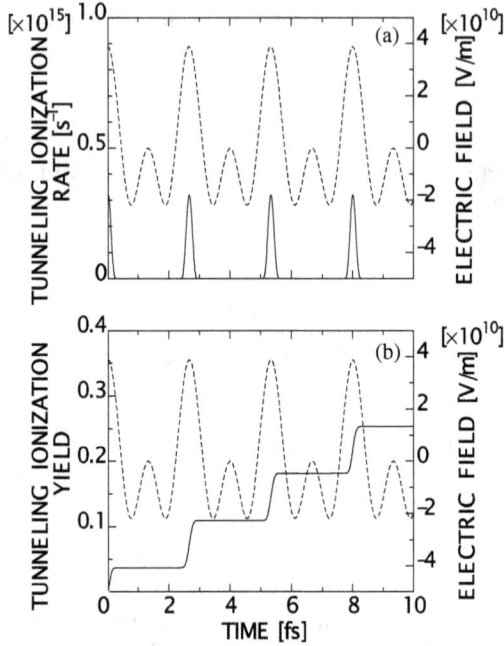

Fig. 9.   Temporal behavior of tunneling ionization for a hydrogen atom (E_{IP}=13.6 eV) in the linearly polarized electric field of a phase-controlled $\omega + 2\omega$ laser pulse. A quasistatic TI model was employed with the following calculation parameters: $\lambda = 400 + 800$ nm, $I = 1.0 \times 10^{14}$ W/cm^2, $F_2/F_1 = 1.0$, $\gamma = 1.07$. (a) Tunneling ionization rate as a function of time; (b) Tunneling ionization yield as a function of time. Dashed curves represent the electric field.

spectra. We consider the quasistatic TI model[41,42,45,46] discussed in Sec. 2.3. By solving the Newton equation of motion for linearly polarized, phase-controlled $\omega + 2\omega$ laser fields with the initial velocity $v(t_0) = 0$, we obtain the following solution for photoelectron velocity $v(t)$:

$$m \frac{dv(t)}{dt} = -eF_1[\cos(\omega t) + r \cos(2\omega t + \phi)],$$

$$v(t) = -\frac{eF_1}{m\omega} \left[\sin(\omega t) + \frac{r}{2} \sin(2\omega t + \phi) \right]$$

$$+ \frac{eF_1}{m\omega} \left[\sin(\omega t_0) + \frac{r}{2} \sin(2\omega t_0 + \phi) \right]. \qquad (11)$$

After the laser pulse passed the interaction region, the first oscillatory term induced by the laser field in Eq. (11) disappears, and the final electron velocity $v_{\text{final}}(t_0)$ and kinetic energy E_k can be expressed by the following equation:

$$v_{\text{final}}(t_0) = \frac{eF_1}{m\omega}\left[\sin(\omega t_0) + \frac{r}{2}\sin(2\omega t_0 + \phi)\right]; \tag{12}$$

$$E_k = \frac{1}{2}mv_{\text{final}}^2$$

$$= 2U_{p,\omega}\sin^2(\omega t_0) + 2U_{p,2\omega}\sin^2(2\omega t_0 + \phi),$$

$$+ 4\sqrt{U_{p,\omega}U_{p,2\omega}}\sin(\omega t_0)\sin(2\omega t_0 + \phi) \tag{13}$$

where $U_{p,\omega}$ and $U_{p,2\omega}$ are the ponderomotive energy for each fundamental and second harmonic light. By summing the TI rates at t_0 that give rise to the same kinetic energy, we can obtain the photoelectron spectra generated by TI as a function of kinetic energy. The relationships between the electric field and final velocity of the photoelectron and corresponding calculated photoelectron spectra for linearly polarized, phase-controlled $\omega + 2\omega$ laser fields at different ϕ are depicted in Fig. 10.

In the case of the relative phase difference $\phi = 0$ (π), the waveform of the electric field is asymmetric so that the photoelectron is seemingly likely to be much more strongly extracted from the negative (positive) side of the atom toward the opposite direction electric field maximum (left panels of Figs. 10(a) and 10(c)). Therefore, we can expect an asymmetric photoelectron spectrum with preferential negative (positive) direction at $\phi = 0$ (π). However, the corresponding calculated photoelectron spectra show symmetric forms (right panels of Figs. 10(a) and 10(c)), because the velocity for the majority of photoelectrons generated at the field maxima is zero (shown by closed circles in left panel of Fig. 10) and the function $v_{\text{final}}(\omega t_0)$ is antisymmetric around $\omega t_0 = 0$ i.e., $v_{\text{final}}(\omega t_0) = -v_{\text{final}}(-\omega t_0)$ with $\phi = 0$ (π).[45,46]

On the other hand, when $\phi = \pi/2$ $(3\pi/2)$, the amplitude of the electric field is the same in the positive and negative directions so that the electron is seemingly equally likely to be extracted from the negative and positive sides of the atom (left panels of Figs. 10(b) and 10(d)). Therefore, we can expect a directionally symmetric photoelectron spectrum at $\phi = \pi/2$ $(3\pi/2)$. However, the corresponding calculated photoelectron spectra show asymmetric forms (right panels of Figs. 10(b) and 10(d)), because the

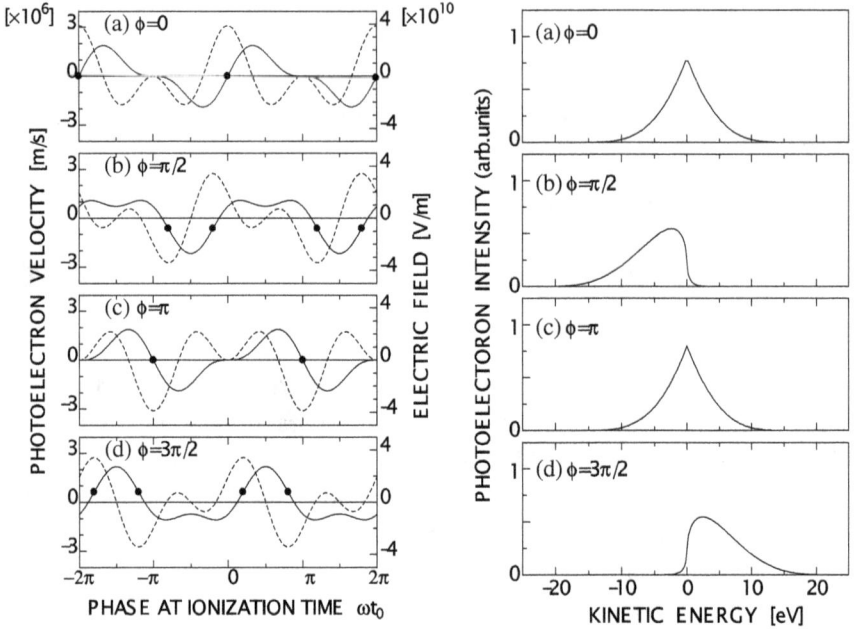

Fig. 10. (Left panel) Velocity of a photoelectron generated by tunneling ionization as a function of phase at ionization time. (Right panel) Corresponding photoelectron spectra as a function of kinetic energy at relative phase differences (a) $\phi = 0$, (b) $\phi = \pi/2$, (c) $\phi = \pi$, and (d) $\phi = 3\pi/2$. A quasistatic TI model employing the same calculation parameters as those listed in the caption for Fig. 9 are used. Dashed curves represent electric fields. Closed circles indicate the photoelectron velocity at electric field maxima. Positive (negative) kinetic energy indicates a photoelectron with positive (negative) velocity.

velocity for the majority of photoelectrons generated at the field maxima are nonzero (negative (positive) values for $\phi = \pi/2$ ($3\pi/2$)) and the function $v_{\text{final}}(\omega t_0)$ for $\phi = \pi/2$ ($3\pi/2$)) is asymmetric.[45,46]

Figure 11 shows the positive/negative yield ratio (I_P/I_N) as a function of ϕ and photoelectron kinetic energy. A clear periodicity of 2π is evident with maximum asymmetry at $\phi = \pi/2$ and $3\pi/2$. The phase-dependent behavior is independent of photoelectron kinetic energy.

To summarize the above discussion, (1) phase-controlled $\omega + 2\omega$ laser fields can be used to manipulate the directionality of TI that occurs in the attosecond time region owing to their highly nonlinear optical response. (2) The quasistatic TI model with phase-controlled $\omega + 2\omega$ laser fields leads to phase-dependent, directionally asymmetric photoelectron spectra with

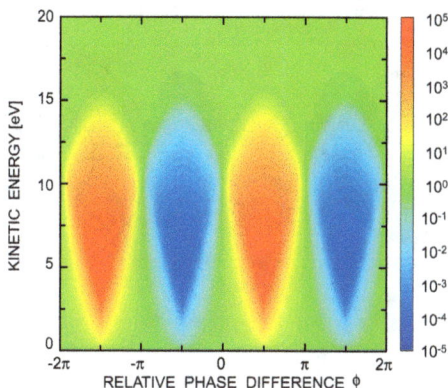

Fig. 11. Density plot of the positive/negative yield ratio (I_P/I_N) for photoelectrons generated by tunneling ionization as a function of the relative phase difference ϕ and photoelectron kinetic energy. A quasistatic TI model employing the same calculation parameters as those listed in the caption for Fig. 9 are used. The I_P/I_N ratios are plotted on a log scale.

maximum asymmetry at $\phi = \pi/2$ and $3\pi/2$. (3) The phase dependence of the photoelectron spectra is independent of the photoelectron kinetic energy. As discussed in Sec. 5.1.2, however, experimental results of phase-dependent behavior are dependent on the photoelectron kinetic energy, and a Coulomb force correction is required for low-energy photoelectrons in the quasistatic TI model.

3.3. *Directionally asymmetric TI (molecules)*

Molecular TI can be described by the molecular ADK model, in which electrons are removed from the HOMO via tunneling.[11-14] According to the molecular ADK model, photoelectrons are much more strongly extracted via tunneling from the large-amplitude lobe of the HOMO along the opposite direction of the electric field vector. Consequently, the angle dependence of the ionization rate reflects the geometric structure of the HOMO. In this section, we consider molecules with asymmetric HOMO structure. Figure 12(a) shows the molecular structure and isocontours of the HOMO of carbon monoxide (CO) determined by *ab initio* calculations using the Gaussian 03W software package.[76]

The HOMO of CO shows an asymmetric σ structure. The angle dependence of the TI rate for CO at a laser intensity of 6×10^{13} W/cm^2

HOMO

(a)

(b)

(c)

(d)

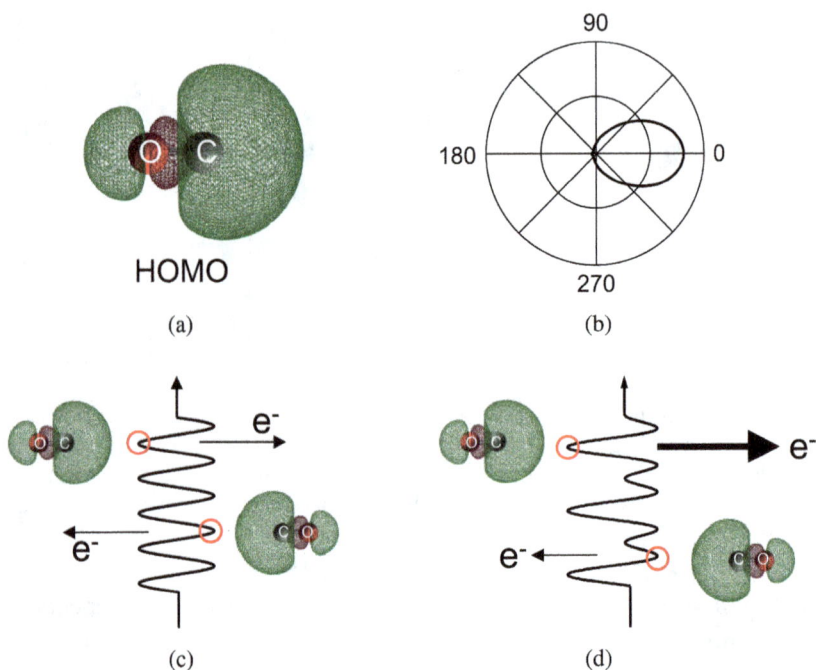

Fig. 12.   (Upper panel) (a) Molecular structure and isocontours of HOMO of the CO molecule. (b) Polar plot of tunneling ionization rate for the CO molecule as a function of angle between laser polarization and molecular axis. The shading indicates the sign of the wavefunction. The calculated data are taken from Ref. 12. (Lower panel) Schematic of the principle of orientation-selective molecular ionization (OSMI). The waveform of the field is shown by the black solid curve. (c) In the case of a single-frequency laser field, molecular tunneling ionization toward positive direction and negative direction has the same ionization rate, resulting in no orientation selectivity. (d) In the case of the phase-controlled $\omega + 2\omega$ laser field, enhanced ionization occurs when the CO molecule is oriented with the electric field maxima pointing toward the O atom, leading to OSMI.

reflecting the geometry of the HOMO has been calculated by using molecular ADK theory (Fig. 12(b)); the calculations show that electrons are much more likely to be removed from the large-amplitude part of the HOMO (carbon) than the small-amplitude part (oxygen).[12] For monochromatic laser fields with a symmetric waveform, however, electrons are removed at the same rate in both the negative direction and the positive direction along the laser polarization so that single-frequency laser fields cannot discriminate the orientation of C-O from that of O-C (lower panel in Fig. 12(c)).

On the other hand, when TI of molecules with an asymmetric HOMO structure is induced by an asymmetric $\omega + 2\omega$ field, electrons are much more likely to be removed from the large-amplitude part of the HOMO in the direction opposite to that of the electric field vector at field maxima, so that enhanced ionization occurs when the CO molecule is oriented with the field maxima pointing toward the O atom (Fig. 12(d)). The $\omega + 2\omega$ laser fields can discriminate among molecular orientation with respect to head–tail ordering, which is impossible to achieve with a single-frequency laser field with a symmetric waveform. Therefore, it has been logically deduced that molecules initially oriented in a certain direction with respect to the asymmetric $\omega + 2\omega$ field are selectively ionized among randomly oriented molecules.

We have experimentally demonstrated for the first time that as a consequence of directionally asymmetric TI of molecules with an asymmetric HOMO, orientation-selective molecular ionization (OSMI) is induced.[57–64] OSMI can be achieved through discrimination of the wavefunction in the space domain by the enhancement of the nonlinear interaction between the asymmetric laser fields and asymmetric HOMO structure. The manipulation of molecular orientation is important for applications such as precision spectroscopy and chemical reactions because orientational averaging, which leads to loss of information or disturbs homogeneous molecular manipulation, can be eliminated.

4. Experimental

The experimental apparatus consisted of laser sources, a phase-controlled $\omega + 2\omega$ laser-field generator,[58,62] and a time-of-flight mass spectrometer (TOF-MS) designed for simultaneous ion–electron detection equipped with a supersonic molecular beam source.[63] The experiments were performed with a Ti:sapphire laser system (Spectra-Physics, Hurricane) operating at 20 Hz or with a Q-switched Nd:YAG laser (Spectra-Physics, LAB-150) operating at 10 Hz. The Ti:sapphire laser system provided pulses of energy at 1 mJ/pulse with a duration of 130 fs at a central wavelength of 800 nm. The Q-switched Nd:YAG laser provided pulses of energy at 500 mJ/pulse with a duration of 10 ns and wavelength of 1064 nm. We inserted a frequency-doubling crystal (β-barium borate (BBO), type-I phase-matching, 1 mm

(a)

(b)

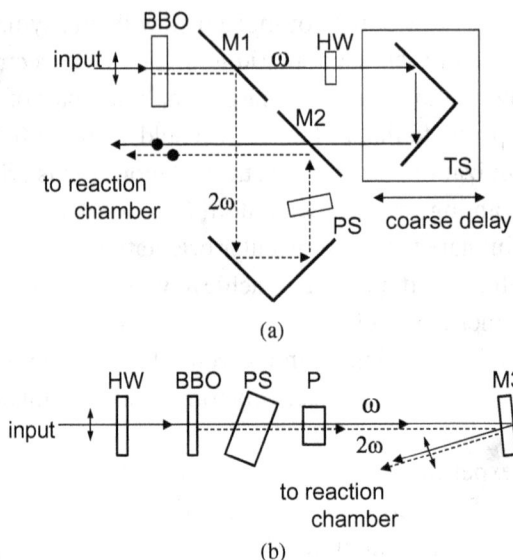

Fig. 13.   Schematic diagrams of the optics used to generate femtosecond phase-controlled $\omega+2\omega$ laser fields: (a) Mach–Zehnder interferometer; (b) robust $\omega+2\omega$ laser field generator. The elements are labeled as follows: BBO: second-harmonic generating crystal; M1, M2: dielectric reflectors for the 2ω beam; HW: half-wave plate; PS: phase shifter (quartz plate); TS: translation stage to control the coarse delay time between the ω and 2ω pulses; P: polarizer; M3: dielectric mirror for propagation delay compensation. The offset between the ω (solid line) and 2ω (dotted line) beams is shown for clarity only, as the two beams overlapped completely in the experiment. The horizontal polarizations are shown as double-headed arrows, and the vertical polarization is shown as closed circles.

(10 mm) thick for femtosecond (nanosecond) pulses, conversion efficiency: 30%) into the path of the laser beam to generate second-harmonic light.

To generate phase-controlled $\omega + 2\omega$ laser pulses, the fundamental light and its second harmonic were introduced into a Mach–Zehnder interferometer.[58] The configuration of the Mach–Zehnder interferometer for femtosecond pulses is shown in Fig. 13(a). The second-harmonic light (dashed line) was separated from the fundamental light (solid line) by a dielectric mirror (M1 in Fig. 13(a)). We inserted a half-wave plate (HW in Fig. 13(a)) that rotated the polarization of the fundamental light by 90° so that the polarizations of the two fields were parallel. The delay time of the two pulses was controlled by a translation stage (TS in Fig. 13(a)) located in the fundamental light path with a resolution of about 4 femtoseconds.

The second-harmonic beams passed through an antireflection coated quartz plate (3 mm thickness) that could be rotated (PS in Fig. 13(a)). This quartz plate was used to change the relative phase difference ϕ of the two fields with a resolution of about 20 attoseconds (0.02π). The ratio of the light intensities (I_2/I_1) was adjusted to be around 0.25 ($F_2/F_1 = 0.5$) by rotating the phase-matching angle of the BBO crystal while keeping the total intensity $I = I_1 + I_2$ constant, where I_1 and I_2 are the intensities of the ω and 2ω pulses, respectively.

Generally, a Mach–Zehnder interferometer is used for generating $\omega + 2\omega$ laser fields. The Mach–Zehnder interferometer is a two-beam interferometer in which a laser beam is separated into two beams along two paths in order to individually adjust their optical phase, and then the two beams are recombined into one; this procedure enables precise handling of the optical phase. However, the most difficult part in using a two-beam interferometer is maintaining the optical phase stability and the spatial overlap of the recombined beams over a long period of time. The optical phase of each beam is individually affected by various fluctuations, such as mechanical vibrations, airflow, and a temperature-dependent distortion of relevant optics mounts over a broad timescale. Furthermore, fine adjustment of the optics is required to recombine the two beams while maintaining their spatial overlap, and a robust phase-controlled $\omega + 2\omega$ laser-field generator that self-compensates any phase fluctuations without fine adjustments is highly desirable.

Figure 13(b) depicts a robust and adjustment-free phase-controlled $\omega + 2\omega$ laser-field generator for femtosecond laser pulses based on a collinear configuration.[62] This optic set does not require interferometric stability because the phase fluctuations in the ω and 2ω beams cancel out when the beams pass through the same path. Furthermore, spatial overlap of the ω and 2ω beams is ensured without optical adjustment because no procedure is required to separate them into the ω and 2ω beams. After the fundamental beam passed through the half-wave plate (HW in Fig. 13(b)) that rotated its polarization direction by $-45°$, second-harmonic pulses polarized to $45°$ were produced by the BBO crystal. Both the fundamental and the second-harmonic pulse passed through a 10 mm-thick phase-shifting quartz plate (PS in Fig. 13(b)) that could be rotated around the incident angle of $45°$ to control the relative phase difference ϕ between the ω and 2ω

pulses. Then, the vertical polarization component of the ω and 2ω pulses was selectively transmitted by a polarizer (an air-spaced Glan–Thompson prism; P in Fig. 13(b)). In the study using femtosecond laser pulses, the difference in propagation delay between the ω and 2ω pulses ($\sim$5 ps) was much larger than their pulse width (130 fs), so that additional optics to achieve temporal overlap of the ω and 2ω beams was required (Fig. 13(b)). The ω and 2ω pulses were then reflected at an angle of $2°$ by a dielectric mirror (0.5 mm thick; M3 in Fig. 13(b)). This mirror has multiple roles: at the front side (99.5% reflectance at 400 nm, 94% transmittance at 800 nm) it reflects the 2ω beam and transmits the ω beam; at the back side (99.5% reflectance at 800 nm, 90% transmittance at 400 nm) it reflects the ω beam and transmits the 2ω beam. The difference in propagation delay between the ω and 2ω pulses generated by all transmitting optics, including the windows of the TOF-MS, were compensated by M3, where the ω beam propagated within the mirror through an additional path not taken by the 2ω beam. By using M3, the robust and adjustment-free $\omega + 2\omega$ laser-field generator could generate femtosecond laser pulses. After being reflected from M3, the phase-controlled $\omega + 2\omega$ beams were directed toward the TOF-MS and were focused on the molecular beam in the TOF-MS with a concave mirror of 120 mm focal length.

Simultaneous ion–electron detection in the TOF-MS is shown schematically in Fig. 14.[63] The supersonic molecular beam of target molecules

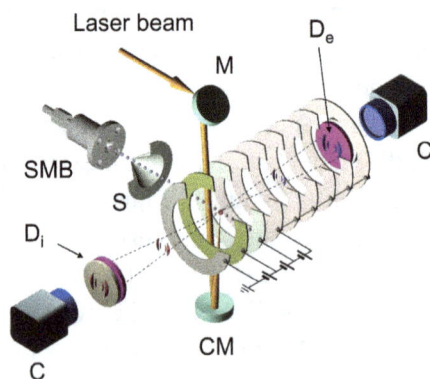

Fig. 14. Schematic of time-of-flight mass spectrometer (TOF-MS) for ion-electron detection. The parts of the spectrometer are labeled as follows: SMB: supersonic molecular beam source; S: skimmer; M: dielectric for the $\omega + 2\omega$ beam; CM: concave mirror for the $\omega + 2\omega$ beam; D_i (D_e): ion (electron) detector; C: CCD camera.

(diluted 5%–10% with helium gas, stagnation pressure: 0.5 MPa, estimated rotational temperature: <20 K) was introduced into the chamber through a pulsed valve (General Valve; 0.5 mm diameter), which was differentially pumped by two molecular turbo pumps (pumping speed: 1800 L/s and 400 L/s). The molecular beam passed through a skimmer (diameter 0.5 mm) located 20 mm from the nozzle. The pressure in the chamber was kept below 3.0×10^{-5} Pa with a 20 Hz repetition rate. After passing through the skimmer, the supersonic molecular beam was ionized by the intense phase-controlled $\omega + 2\omega$ beam.

The TOF-MS mainly consists of a Wiley–McLaren type two-stage accelerator, field-free drift regions for electrons and ions, and two opposing position-sensitive detectors.[63] Photofragment ions (photoelectrons) generated by the $\omega + 2\omega$ pulses are accelerated down (up) by static electric fields toward the opposing detectors. The electrode rings used for acceleration incorporate an electrostatic lens to improve the momentum resolution.[81] After passing through a drift tube at an applied voltage of 0 kV for photofragment ions (2.0 kV for photoelectrons), the photofragment ions (photoelectrons) are detected by a position-sensitive detector composed of a microchannel plate (MCP) with a phosphor screen (77 mm diameter) that was employed to measure both the arrival time and the position of the photofragment ions and photoelectrons at the MCP detector. One-dimensional (1D) TOF spectra for photofragment ions were recorded by a digital oscilloscope. Two-dimensional angular distributions of photofragment ions (photoelectrons) that converted on the position-sensitive detector were recorded by a CCD camera system. Mass selectivity of the fragment ions for the two-dimensional (2D) images was achieved by gating the gain of the detector (temporal width: 100 ns) at the arrival time of each photofragment ion. In this configuration, we could simultaneously measure the phase-dependence of both the photofragment ions and the photoelectrons under identical conditions of the relative phase difference ϕ and laser intensity. The energy resolution of the ion detector (photoelectron detector) was estimated to be 0.1 eV. Our TOF-MS for ion–electron detection is similar to cold ion target recoil ion momentum spectroscopy (COLTRIMS) without a magnetic field.[81] However, since our method does not require coincidence measurements, it is applicable to molecular systems for which coincidence measurements are difficult to perform.

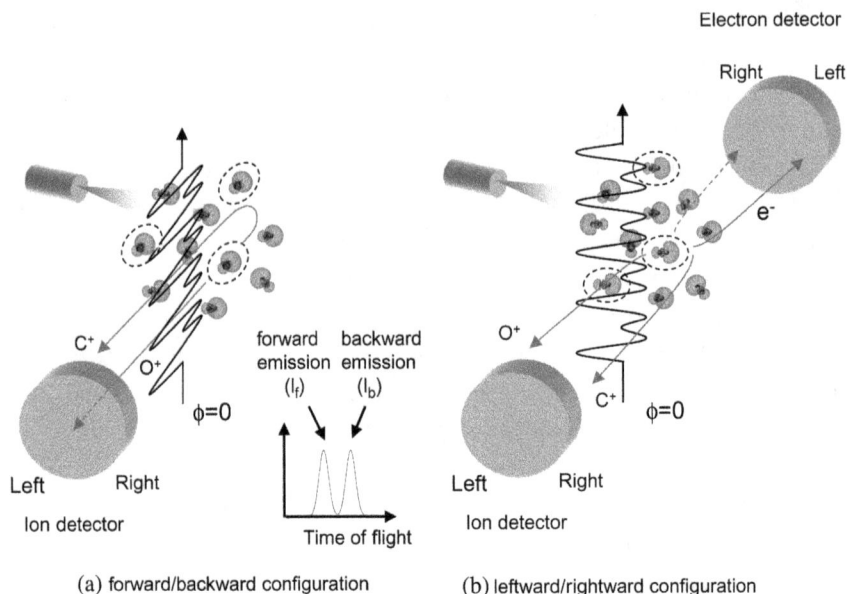

Fig. 15. Schematic of the experimental configuration for simultaneous ion–electron detection in directionally asymmetric molecular tunneling ionization. The waveform of a phase-controlled two-color $\omega + 2\omega$ laser field at a relative phase difference of $\phi = 0$ is shown by the black solid curve. Trajectories of photofragment ions and photoelectrons are shown by the curved gray arrows. (a) forward/backward configuration; (b) leftward/rightward configuration.

We define the experimental configuration among the polarization direction, the detection axis, and the direction of electric field maxima at relative phase difference $\phi = 0$ (Fig. 15). In the measurement of 1D TOF spectra, the polarization direction of the $\omega + 2\omega$ laser fields is set to be horizontal and parallel to the detection axis, and we define $\phi = 0$ when the electric field maxima points toward the ion detector (forward/backward configuration; Fig. 15(a)). For the measurement of the 2D photofragment (photoelectron) angular distribution, the polarization direction of the $\omega + 2\omega$ laser fields is set to be horizontal and perpendicular to the detection axis, and we define $\phi = 0$ when the electric field maxima points leftward (rightward) with respect to the ion (electron) detector (leftward/rightward configuration; Fig. 15(b)).

To calibrate ϕ, at first we simultaneously measured dissociative ionization in target molecules and the optical interference between 2ω

beams generated in the first crystal and second frequency-doubling crystal in the vacuum chamber.[41,42] This procedure, however, was difficult to perform because it involved handling of the optical equipment in the vacuum chamber. Alternatively, we performed a simultaneous measurement using gas mixtures of target molecules and reference molecules.[60] This method provides an accurate phase relationship between target molecules and reference molecules under identical experimental conditions, i.e., within the same experimental run.

5. Results and Discussion

5.1. *Diatomic molecule: CO*

5.1.1. *Photofragment detection*

Figure 16 shows the TOF mass spectrum of ions when CO molecules (ionization potential $E_{IP} = 14.0\,eV$) were irradiated with $\omega + 2\omega$ pulses in the forward/backward configuration. We estimated the total intensity $I = I_1 + I_2$ to be approximately $5 \times 10^{13}\,W/cm^2$ ($I_1 = 4 \times 10^{13}$, $I_2 = 1 \times 10^{13}\,W/cm^2$) at the focus. The corresponding Keldysh adiabatic parameter was $\gamma = 1.5$. The generation of singly charged parent CO^+ is the main process (>95%; Fig. 16, inset). Expanded views of the spectra from singly and doubly photofragmented ions show a pair of peaks, one corresponding to emission directly toward the detector, and the other corresponding to ejection in the backward direction before reversal by the extraction fields (Fig. 15(a)). The assignment of each dissociation channel has been reported as a Coulomb explosion process $CO^{+(p+q)} \rightarrow C^{+p} + O^{+q}$ (where p and q are integers).[82] Strong forward/backward asymmetries show that the C^+ and C^{2+} (O^+ and O^{2+}) ions were preferentially emitted away from (toward) the detector at $\phi = 0$, when the electric field maximum pointed toward the detector. Conversely, the directional asymmetries of each of the photofragments were reversed at $\phi = \pi$. Corresponding 2D angular distributions of the photofragment ions with pronounced angular localization in the leftward/rightward configuration show that a prominent degree of selectivity was achieved both in the orientation direction and in the angular distribution (Fig. 16, images).

A clear periodicity of 2π was observed in the leftward/rightward yield ratio (I_L/I_R) as a function of ϕ for all photofragments displayed (Fig. 18).

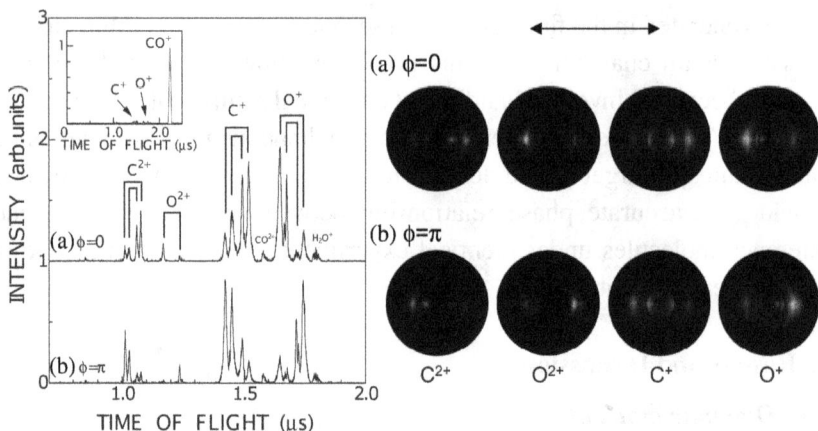

Fig. 16. Graphs: TOF mass spectra of ions generated by the dissociative ionization of CO molecules irradiated with the phase-controlled $\omega + 2\omega$ laser fields in the forward/backward configuration (inset). Expanded TOF mass spectra of photofragment ions at relative phase differences (a) $\phi = 0$ and (b) $\phi = \pi$. The solid lines indicate pairs of forward and backward peaks. Images: Angular distributions of photofragment emission generated by the dissociative ionization of CO molecules irradiated with phase-controlled $\omega + 2\omega$ laser fields in the leftward/rightward configuration at relative phase differences (a) $\phi = 0$ and (b) $\phi = \pi$. The double-headed arrow indicates the direction of polarization. The data are taken from Ref. 63.

The phase dependence between $C^+(C^{2+})$ and $O^+(O^{2+})$ are completely out of phase with each other. This result shows that phase-controlled $\omega + 2\omega$ pulses can discriminate the molecular orientation of head–tail order. Furthermore, the phase dependence between $C^+(O^+)$ and $C^{2+}(O^{2+})$ shows completely in-phase behavior, indicating that the direction of molecular orientation for singly and doubly charged CO are the same.

The photofragment behavior observed (Figs. 16 and 18) suggests that the direction of the detected molecules was consistent with that expected by the molecular ADK model described in Sec. 3.3; i.e., electrons are much more strongly removed by tunneling from the large-amplitude part (carbon) of the HOMO opposite to the direction of the electric field vector at its maxima for $\phi = 0$ and $\phi = \pi$. These results are in good agreement with the recently published experimental result of orientation-dependent TI of CO molecules induced by single-color circularly and elliptically polarized femtosecond laser pulses with an intensity of 4.0×10^{14} W/cm^2 and a pulse duration of 35 fs in COLTRIMS spectroscopy.[83] The Stark-corrected

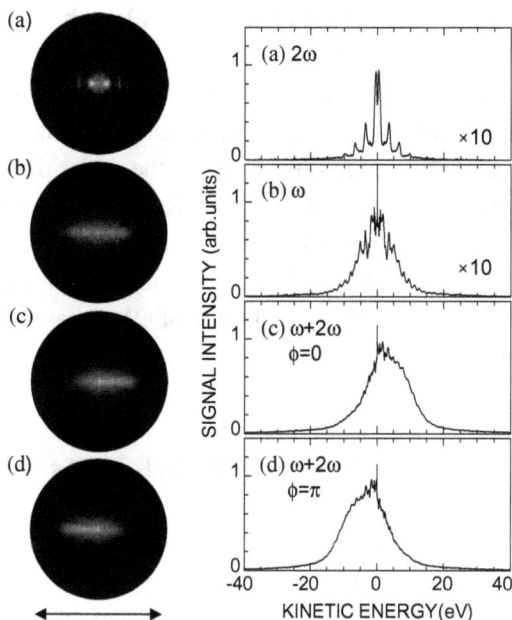

Fig. 17. Images: Angular distributions of photoelectron emission generated by the irradiation of (a) the second harmonic (2ω) lights, (b) the fundamental (ω) lights, and the phase-controlled two-color $\omega + 2\omega$ laser fields at relative phase differences (c) $\phi = 0$ and (d) $\phi = \pi$. Graphs: Photoelectron spectra as a function of kinetic energy along the polarization direction (double-headed arrow) converted from respective images. Positive (negative) kinetic energy corresponds to rightward (leftward) photoelectron emission. The data are taken from Ref. 63.

molecular ADK model has predicted an orientation-dependent ionization that is opposite to the traditional molecular ADK model in the case of CO molecules [77–80]; therefore, the linear Stark effect plays a minor role in the TI for CO molecules.[83]

There is another conceivable mechanism related to the detection of oriented molecules: dynamic molecular orientation (DMO).[50–56] Several theoretical investigations have reported that molecules can be dynamically oriented along the laser polarization direction by the torque generated by the nonlinear interaction between a nonresonant $\omega + 2\omega$ laser field and the permanent dipole[50,51] or hyperpolarizability of molecules[52,54] (the linear interaction between an $\omega + 2\omega$ laser field and the permanent dipole of molecules averages to zero over an optical cycle). If the laser pulse is longer than the rotational period of the molecules, then molecules orient adiabatically during laser irradiation (adiabatic molecular orientation).[52,53] If the laser pulse is shorter than the rotational period of the molecules, then rotational wave packets are formed and dynamical orientation is reconstructed at revival times even after the laser irradiation ceases (nonadiabatic molecular orientation).[50,51] Recently, DMO based on the hyperpolarizability of molecules has been achieved both adiabatically[56] and nonadiabatically.[55] However, the degrees of orientation have been observed to be very small. Therefore, the contribution of DMO based on the hyperpolarizability of molecules during the pulse duration in our experiments can be neglected.

We note two controversial points. First, we are aware of the controversy concerning the boundary between MPI and TI. As indicated earlier, Keldysh theory states that if $\gamma > 1$, MPI is dominant.[4] Since there is no absolute boundary between MPI and TI, in the intermediate region $\gamma \sim 1$, phenomena can often be successfully explained by both MPI and TI. Our experimental conditions correspond to $\gamma = 1.5$, where MPI is considered to be dominant. Nonetheless, in terms of directionally asymmetric TI, the expression can be useful for intuitive understanding of our experimental results. The intermediate regime $\gamma \sim 1$ is referred to as nonadiabatic TI.[9] The theory concerning this nonadiabatic model has been used to point out that TI is valid even for $\gamma \gg 1$. Some relevant investigations concerning the boundary between MPI and TI are as follows: (i) Uiberacker *et al.* have reported real-time observation of the optical TI process by using an attosecond XUV pulse pumping and NIR few-cycle pulse probing technique, and they have shown that TI remains the dominant ionization mechanism even at $\gamma \sim 3$,[74]; (ii) Nonadiabatic TI in helium with γ ranging from 1.17 to 1.45 has been experimentally explored by means of an attosecond angular streaking method, and the authors of that study concluded that the TI model was valid for those conditions[75]; (iii) Dewitt and Levis have observed that a transition from MPI to TI occurs in polyatomic molecules by changing the electron delocalization through the molecular structure, and they have shown that a large electronic orbital size reduces γ effectively. In other words, TI can be dominant even for $\gamma > 1$;[84] (iv) Reiss has pointed out that there are disqualifying features in categorization using γ where ionization with $\gamma \gg 1$ can occur only by TI, and ionization with $\gamma \ll 1$ must be induced by a more intense laser regime such as the over-the-barrier process.[85]

Second, there exists some uncertainty about monitoring the ionization process through dissociative ionization channels. Dissociative ionization processes include several entangled non-sequential processes such as (1) the generation of the parent ion in dissociative states; (2) recollision-induced electron excitation; and (3) direct generation of the parent ion through an electron ejection from the next lower electronic state HOMO-1,[83,85-87] which is likely to be more prominent for experiments with higher laser intensity. These processes might have induced deviation from the molecular ADK model. Considering the laser intensity in the experiment shown in Figs. 16 and 17, these non-sequential processes should have certain contributions to the TOF spectra. Despite these situations, the experimental results observed here for CO suggest that the direction of the detected molecules was consistent with that expected by the molecular ADK model. Therefore, from an experimental viewpoint, TI based on the molecular ADK model seems to be the main process that occurred in CO, and any dissociative ionization processes that could have induced deviations from the molecular ADK model seem to have occurred to a much smaller extent, if at all. The orientation-dependent TI rate of CO molecules induced by single-color laser pulses with an intensity of 4.0×10^{14} W/cm^2 have been measured by COLTRIMS spectroscopy.[83] The contribution of HOMO-1 to the dissociative single ionization process in that experiment was $\sim$30% of the total signal in the experiment, with a laser intensity of about 10 times higher than that used in our experiments. Therefore, we can safely say, that TI based on the molecular ADK model is the primary step that overcome other effects followed by TI process.

5.1.2. *Photoelectron detection*

Figure 18 shows the photoelectron spectra of ions when CO molecules were irradiated with femtosecond laser pulses in the leftward/rightward configuration. The 2D angular distribution of the photoelectrons under irradiation from only the 2ω pulse in the leftward/rightward configuration was observed as a series of clear discrete symmetric ring structures localized in the polarization direction (Fig. 18(a), left column). In the corresponding photoelectron spectrum as a function of kinetic energy (Fig. 18(a), right

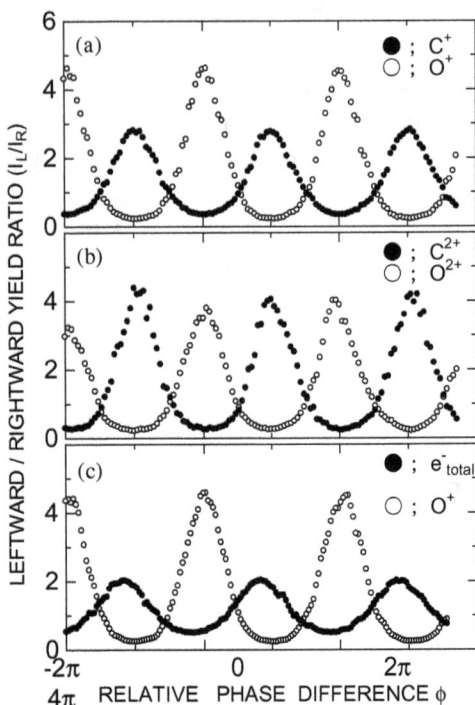

Fig. 18. Leftward/rightward yield ratio (I_L/I_R) of (a) singly charged and (b) doubly charged photofragment ions as a function of relative phase phase difference ϕ: (open circles) oxygen; (closed circles) carbon. (c) I_L/I_R as a function of relative phase difference ϕ: (open circles) oxygen; (closed circles) total photoelectrons in simultaneous ion–electron detection. The data are taken from Ref. 63.

column), the energy spacing of the series of peaks is 3.1 eV, which corresponds to the photon energy of the 2ω pulse. This pattern results from the well known ATI, where the lowest peak corresponds to MPI by overcoming the ionization potential ($E_{IP} = 14.0\,\text{eV}$), and the subsequent peaks correspond to the absorption of additional photons.

The 2D angular distribution of photoelectrons under irradiation from only the ω pulse was observed as a strongly localized symmetric angular distribution reflecting the polarization of the ω pulse, accompanied by a faint series of discrete ring structures (Fig. 18(b), left column). In the corresponding photoelectron spectrum, the energy distribution shows a broad and exponentially decreasing dependence superimposed on a weak series of discrete peaks whose energy spacing is 1.55 eV (half the 2ω irradiation), indicating the transition from MPI to TI.[8]

When CO molecules were irradiated by the phase-controlled $\omega + 2\omega$ pulses, the intensity of the photoelectron signal became about 10 times the sum of each signal for the ω and the 2ω irradiation due to the highly nonlinear optical process. Figures 18(c) and 18(d) show the 2D angular distribution of photoelectrons and the corresponding photoelectron spectra at $\phi = 0$ and π, respectively. The disappearance of the discrete structures in the photoelectron spectra indicates that the laser intensity reached the TI regime. Furthermore, strong directional asymmetry in the leftward/rightward emission is clearly observed. This asymmetry shows that the photoelectrons were preferentially emitted rightward (leftward) of the electron detector at $\phi = 0$ (π).

Most importantly, the quantum dynamics of photoelectrons can be tracked by using the phase-dependent oriented molecules as a phase reference in simultaneous ion–electron detection. If the photoelectrons are removed via tunneling from the large amplitude lobes of the HOMO opposite to the maxima of the electric fields (hereafter called "intuitive" photoelectron emission, following Refs. 45 and 46; solid gray line of the photoelectron orbit in Fig. 15(b)), the leftward/rightward asymmetry between O^+ and the photoelectrons is expected to exhibit in-phase behavior. Figure 17(c) shows the I_L/I_R ratio as a function of ϕ in the simultaneous measurement of O^+ and total photoelectrons. The O^+ and photoelectrons are nearly out of phase with each other (the phase lag with respect to O^+ is $0.85\,\pi$). Experimental results show that the photoelectrons are emitted nearly opposite to the intuitive direction. Figure 19(a) shows a density plot of the I_L/I_R ratio as a function of ϕ and photoelectron kinetic energy in the simultaneous measurement of O^+ and photoelectrons. The phase-dependent behavior was dependent on photoelectron kinetic energy, and can be divided into two regions: photoelectrons with low kinetic energy (0–0.3 a.u.) with directional asymmetry around $\phi = 0$ (π), and photoelectrons with high kinetic energy (0.3–0.7 a.u.) with directional asymmetry around $\phi = \pi/2$ ($3\pi/2$).

We now discuss the quantum dynamics of photoelectrons generated by $\omega + 2\omega$ laser fields. In early studies of photoelectron dynamics generated by irradiation of molecules with intense $\omega + 2\omega$ laser fields, the

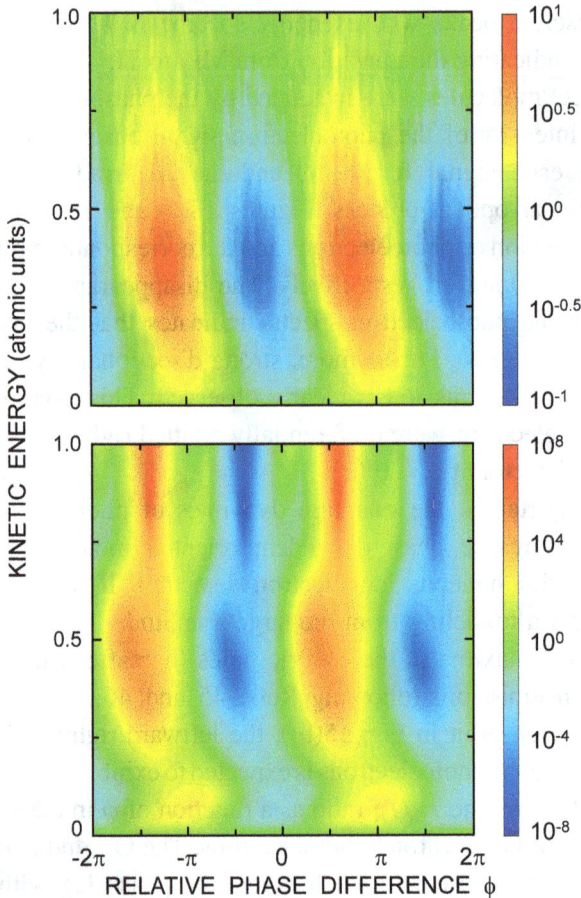

Fig. 19. Density plot of the leftward/rightward yield ratio (I_L/I_R) for photoelectrons as a function of the relative phase difference ϕ and photoelectron kinetic energy (atomic units): (a) experimental result, (b) numerical calculation. Note that the I_L/I_R ratios are plotted on a log scale. The data are taken from Ref. 63.

puzzling behavior of directionally asymmetric emission between positively charged nuclear fragments and photoelectrons was observed.[38] Bandrauk and Chelkowski have discussed theoretically the details of directionally asymmetric photoelectron emission induced by $\omega + 2\omega$ fields.[45,46] First, although the quasistatic TI model with $\omega + 2\omega$ laser fields (described in Sec. 3.2) predicts no directional asymmetry at $\phi = 0$ (π), it does predict directional asymmetry at $\phi = \pi/2$ ($3\pi/2$),[41,42,45,46] so the photoelectrons with high kinetic energy in our experiment can be explained by the quasistatic TI model. Second, Bandrauk and Chelkowski calculated the numerical solution of the time-dependent Schrödinger equation (TDSE) for 1D H_2^+ molecules and H atoms in phase-controlled $\omega + 2\omega$ laser fields, and found that photoelectrons are emitted opposite to the intuitive direction at $\phi = 0$ (π). They used a Coulomb-corrected quasistatic TI model to explain that the origin of the counterintuitive photoelectron emission is Coulomb attraction from the parent ion.[45,46] This Coulomb-corrected quasistatic TI model adequately explains our experimental results for low kinetic energy photoelectrons.

The author and coworker have performed a numerical calculation of the 3D TDSE.[88] In brief, we considered a hydrogen atom interacting with a $\omega + 2\omega$ laser field with a pulse-duration of 10 fs, in which asymmetric photoelectron emission induced by the Carrier-envelope phase was negligible. The total intensity of the $\omega + 2\omega$ laser field $I = I_1 + I_2$ in the calculation was set to be 5×10^{13} W/cm^2 ($I_1 = 4.0 \times 10^{13}$, $I_2 = 1.0 \times 10^{13}$ W/cm^2). To smooth out the ATI peaks, the ratio was obtained by averaging the calculated spectra over bins of $\Delta p = 1.8$ eV. In Fig. 19, the experimental results and the numerical calculation are compared. Our experimental results are qualitatively in agreement with the numerical results obtained by the 3D TDSE, although the absolute value of the I_L/I_R ratio in the experimental results is smaller than that in the numerical calculation mainly owing to experimentally imperfect conditions such as spatial overlapping between the ω and 2ω laser fields. The experimental results might fail to detect the fine structure seen in the numerical calculation for fast photoelectrons greater than 0.7 a.u. because of the low sensitivity of the electron detector. We can interpret the quantum dynamics of photoelectrons generated by the $\omega + 2\omega$ field by 3D TDSE by considering the previously reported quasistatic TI model and Coulomb-corrected quasistatic TI model. First, the phase-dependent behavior of the photoelectrons with low kinetic energy is in good agreement with the Coulomb-corrected quasistatic model. This result can be explained by the effect in which photoelectrons with low kinetic energy are emitted toward the counterintuitive direction owing to Coulomb attraction from the parent ion.[45,46] Second, the phase-dependent behavior of the photoelectrons with high kinetic energy asymptotically approaches the quasistatic TI model, which predicts directional asymmetry at $\phi =$

$\pi/2$ $(3\pi/2)$.[21,28,29] This result is consistent with the physical situation because photoelectrons with high kinetic energy, which are less affected by Coulomb interaction than those with low kinetic energy, are driven by the intense $\omega + 2\omega$ laser fields, overcoming the Coulomb attraction. We have successfully observed the transition from slow photoelectrons to fast photoelectrons in the phase-dependent behavior of directionally asymmetric photoelectron emission induced by the $\omega + 2\omega$ laser field. Finally, the fine structure for fast photoelectrons greater than 0.7 a.u. in the numerical calculation includes the backscattering of photoelectrons by parent ions.[49] Further experimental studies with highly sensitive photoelectron detection are required to examine the very sensitive phase-dependent behavior of the backscattered photoelectrons.

5.2. *Other molecules*

To confirm that the main mechanism of the detection of oriented molecules is OSMI, we investigated the dissociative ionization of molecules induced by $\omega + 2\omega$ laser fields with pulse durations of 130 fs and 10 ns by changing the parameters of the molecules systematically, as shown in Figs. 20(a)–20(d).

5.2.1. *Nonpolar molecule with asymmetric structure: Br(CH$_2$)$_2$ Cl*

As an example of a nonpolar molecule with an asymmetric structure, we have chosen 1-bromo-2-chloroethane (BCE) ($E_{IP} = 10.55$ eV).[59] Figure 20(a) shows the molecular structure and HOMO of BCE as determined by *ab initio* calculations using the Gaussian 03W software package (method: MP2; basis set: 6-311+G(d,p)).[76] Among the three possible rotational isomers, the trans isomer shown in Fig. 20(a) is the most stable in the gas phase. The BCE molecule has a very small permanent dipole moment (calculated value: 0.0057 Debye, pointing from Cl to Br) due to cancellation of two halogen atoms with large electronegativities (Cl: 3.0; Br: 2.8) located on opposite sides of the molecule. However, the HOMO shows a π structure with large asymmetry along the molecular frame (Fig. 20(a)).

The experiment was performed with a laser intensity (1.0×10^{13} W/cm^2) in the vicinity of the regime, where doubly charged fragment ions due to Coulomb explosion were observed.

(a) nonpolar molecule (Br(CH$_2$)$_2$Cl)

(c) systematically changing molecular system (CH$_3$X; X=F, Cl, Br, I)

(b) large molecule(C$_6$H$_{13}$I)

CH$_3$F
$\mu = 1.86$ D

CH$_3$Cl
$\mu = 1.90$ D

CH$_3$Br
$\mu = 1.82$ D

(d) OCS molecule investigated by phase-controlled ($\omega+2\omega$) nanosecond laser pulse

$\mu = 0.72$ D

CH$_3$I
$\mu = 1.640$D

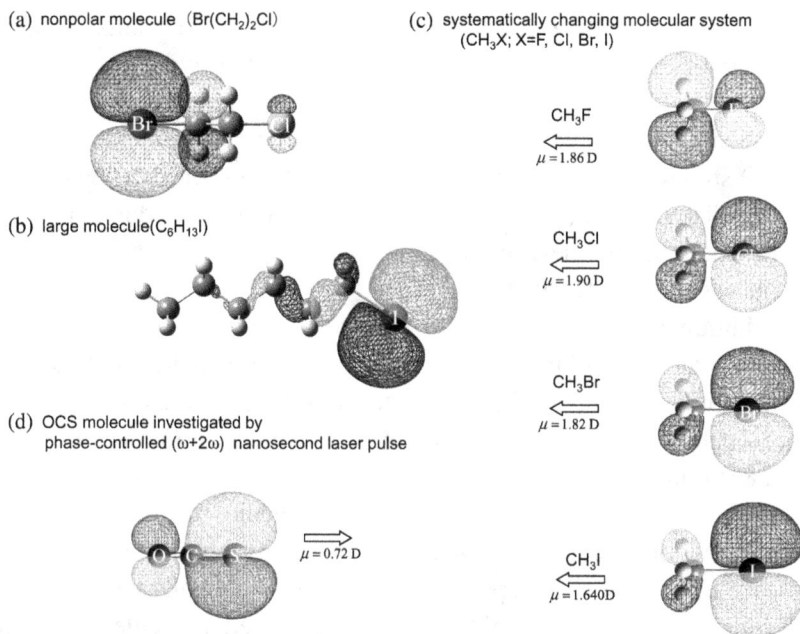

Fig. 20. Molecular structures and isocontours of the HOMO of investigated molecules as determined by *ab initio* calculations using the Gaussian 03W software package. Shadings indicate the signs of the wavefunctions. The directions of the permanent dipoles are shown by thick arrows.

When BCE molecules were irradiated with $\omega + 2\omega$ laser pulses in the forward/backward configuration, various singly charged photofragment ions and parent ions were detected in the TOF mass spectrum. Directional asymmetries in the forward/backward emissions were observed in various photofragment ions, and a clear periodicity of 2π was observed in the I_f/I_b ratio for all photofragments. The phase dependencies between the Cl$^+$ (Br$^+$) ions and counter cations were completely out of phase with each other. This result shows that the phase-controlled $\omega + 2\omega$ fields can discriminate molecular orientation from head–tail order. Moreover, the phase dependencies between Cl$^+$ and Br$^+$ were also out of phase with each other. It is evident from all phase dependencies that phase-controlled $\omega+2\omega$ fields discriminate the molecular orientation of the head–tail order.[59] Br$^+$ ions were preferentially emitted away from the detector and Cl$^+$ ions were preferentially emitted toward the detector at $\phi = 0$, when the electric field maxima pointed toward the detector. These observed results show that the direction of the detected molecules is consistent with that expected by the molecular ADK model. Therefore, we conclude that the phase-controlled $\omega + 2\omega$ field achieves OSMI, reflecting the asymmetry of the HOMO structure.[59] Even for nonpolar molecules, OSMI can be achieved through discrimination of the wavefunction in the space domain by the enhancement of nonlinear interaction between the asymmetric laser fields and the asymmetric HOMO structure.

Additionally, we mention the relative angle between the oriented molecules and the polarization direction of the laser fields, and the contribution of the induced dipole moment. Alnaser *et al.* observed a butterfly-shaped pattern reflecting the structure of π orbitals in O_2 molecules by using 8 fs optical pulses and 2D photofragment-emission pattern imaging, where the direction of selectively ionized molecules was 40° relative to the polarization direction.[13] When 35 fs pulses were used instead of 8 fs pulses, the butterfly-shaped pattern changed to a dumbbell-shaped pattern, indicating that the direction of the selectively ionized molecules was along the direction of the laser polarization due to dynamic alignment (not orientation) by the induced dipole during the laser pulse.[13,89] Almost all molecules experience some contribution of dynamic molecular alignment due to an induced dipole. Thus, it is possible that our 130 fs $\omega + 2\omega$ pulse induces dynamic alignment, even for relatively heavy BCE molecules, and that our measurement is a result of the OSMI in aligned molecules, rather than in randomly oriented molecules, during the laser pulse.

5.2.2. *Large molecule: $C_6H_{13}I$*

We studied a large polyatomic molecule, 1-iodohexane ($C_6H_{13}I$) ($E_{IP} = 9.20$ eV), to determine whether OSMI could be achieved.[62] The molecular structure and isocontours of the HOMO of $C_6H_{13}I$ (Fig. 20(b)) were determined by *ab initio* calculations using the Gaussian 03W software package (method: MP2; basis sets: LanL2DZ augmented by polarization functions and diffuse functions).[76] The HOMO of $C_6H_{13}I$ was remarkably asymmetric due to "squeezing" by the iodine atom (Fig. 20(b)). Although iodination is a simple chemical treatment, it can induce a dramatic change at the wavefunction level and can therefore be used to "quantum mark" molecules when designing wavefunctions.

We adjusted the laser intensity (2.0×10^{13} W/cm^2) to be near the regime where doubly charged fragment ions due to Coulomb explosions were observed (a small I^{2+} signal was observed).

When $C_6H_{13}I$ molecules were irradiated with femtosecond $\omega + 2\omega$ pulses in the forward/backward configuration, various singly charged photofragment ions such as hydrocarbon cations and iodine-containing cations, as well as parent ions, were detected in the TOF mass spectrum. Directional asymmetries in the forward/backward emissions were observed for various photofragment ions, and a clear periodicity of 2π was observed in the I_f/I_b ratio for all photofragments except $C_6H_{13}^+$. The phase dependencies of iodine and iodine-containing cations were completely out of phase with those for the carbon and hydrocarbon cations. This result shows that a phase-controlled $\omega + 2\omega$ optical field discriminates the head–tail order of molecules.[62]

We draw two direct conclusions from these experimental results. First, the phase dependencies of all photofragments except $C_6H_{13}^+$ were consistent with the molecular structure of 1-iodohexane. Therefore, we can reasonably conclude that the prompt axial recoil approximation is valid even for large polyatomic molecules, and that the photofragment emission pattern reflects the molecular structure. Regarding $C_6H_{13}^+$, there are two conceivable explanations for the fact that the $C_6H_{13}^+$ fragments did not show a phase-dependent behavior: (1) a slow dissociation process, in which $C_6H_{13}^+$ was produced on a time scale longer than the rotational period, allowing orientation averaging; and (2) a kinetic energy that was too low to show a photofragment emission pattern. The excess energy that molecular cations obtain during the ionization process is divided between the translational and internal (vibrational and rotational) energy of the photofragments. The translational energy of the photofragments decreases with increasing number of constituent atoms, because the number of internal degrees of freedom increases. Although the observed phase difference decreased with increasing number of constituent atoms, favoring explanation (2), at present we cannot entirely exclude either explanation.

Second, our results indicate that the direction of the detected molecules was consistent with that expected by the molecular ADK model. Therefore, it is reasonable to conclude that the molecular ADK model is valid for OSMI even for large polyatomic molecules.[62] OSMI is free of the constraints of size and weight of molecules, and this is an advantage compared to DMO, with which it is difficult to orient large heavy molecules that require large torques at practical laser intensities.

5.2.3. *Systematically changing molecular system:* $CH_3X(X=F, Cl, Br, I)$

We have also investigated the phase-sensitive ionization related to molecular orientation induced by intense phase-controlled $\omega + 2\omega$ pulses in the case of systematically changing orbital asymmetry.[60] Figure 20(c) shows the molecular structures and isocontours of the HOMO of four methyl halide molecules (CH$_3$F: E_{IP} = 12.47 eV, CH$_3$Cl: E_{IP} = 11.22 eV, CH$_3$Br: E_{IP} = 10.54 eV, CH$_3$I: E_{IP} = 9.54 eV) as determined by *ab initio* calculations using the Gaussian 03W software package[76] (method: MP2; basis sets: 6-31++G(2df,p) for CH$_3$F and CH$_3$Cl, and LanL2DZ augmented by polarization functions and diffuse functions for CH$_3$Br and CH$_3$I). The HOMO of all the methyl halide molecules shows an asymmetric π structure, and the degree of asymmetry changes systematically with respect to the halogen atom. The wavefunctions for the halogen-atom side are larger than that of the methyl moeities in CH$_3$I, CH$_3$Br, and CH$_3$Cl. The degree of asymmetry decreases gradually from CH$_3$I to CH$_3$Cl, and then reverses for CH$_3$F. Thus, if OSMI based on the molecular ADK model is the main orientation process, the orientation direction of selectively ionized CH$_3$F is opposite of that for CH$_3$Cl, CH$_3$Br, and CH$_3$I.

We adjusted the laser intensity to near the regime where doubly-charged fragment ions due to Coulomb explosions were observed. The minimum intensity was 10^{12} W/cm^2 for CH$_3$I and the maximum intensity was 10^{13} W/cm^2 for CH$_3$F, reflecting the difference in ionization potentials among the molecules.

When methyl halide molecules were irradiated with femtosecond $\omega + 2\omega$ pulses in the forward/backward configuration, various singly charged photofragment ions and parent ions were detected in the TOF mass spectrum. The directional asymmetries in the forward–backward emissions were observed in various photofragment ions, and a clear periodicity of 2π was observed in the I_f/I_b ratio for all photofragments. The phase dependencies between the halogen ions and the CH_3^+ cations were completely out of phase with each other for CH_3I, and approximately out of phase with each other for CH_3Br, CH_3Cl, and CH_3F. This result shows that a phase-controlled $\omega + 2\omega$ laser field discriminates the head–tail order of oriented molecules.[60]

To classify the direction of the oriented molecules, we performed a simultaneous measurement using gas mixtures of CH_3I/CH_3Br, CH_3I/CH_3Cl, and CH_3I/CH_3F.[60] The I^+, Br^+, and Cl^+ ions exhibited completely in-phase behavior, while the F^+ ion was approximately out of phase with the other three halogen atoms. This result indicates that the direction of oriented molecules is the same in CH_3I, CH_3Br, and CH_3Cl, and that the CH_3F molecule is oriented in the opposite direction from the other three methyl halides. The classification by phase behavior is consistent with that expected by OSMI based on the molecular ADK model. Moreover, the directions of the detected molecules are consistent with those expected by OSMI, whereas the large-amplitude parts (halogen atoms for CH_3I, CH_3Br, and CH_3Cl, and the methyl moieties for CH_3F) were located on the backward side and ionized electrons were removed backward at $\phi = 0$ when the optical electric field maximum pointed toward the detector. Therefore, it is reasonable to conclude that OSMI based on the molecular ADK model is the main process occurring in the phase-sensitive ionization of the four methyl halides induced by a phase-controlled $\omega + 2\omega$ field.[60]

5.2.4. *OCS molecule investigated by nanosecond $\omega + 2\omega$ laser fields*

Finally, we investigated the dependence of laser pulse duration.[61] We investigated OCS molecules ($E_{IP} = 11.18\,eV$) by nanosecond phase-controlled $\omega + 2\omega$ ($\lambda = 1064 + 532\,nm$) pulses generated by the Nd:YAG laser with an intensity of $5.0 \times 10^{12}\,W/cm^2$ and a pulse duration of $10\,ns$.

When OCS molecules were irradiated with nanosecond $\omega + 2\omega$ pulses in the forward/backward configuration, singly charged OC^+, S^+, and parent OCS^+ were detected in the TOF mass spectrum. Forward/backward asymmetry was clearly observed in the TOF spectrum. The forward peak of the OC^+ ions was more dominant than the backward peak, and the backward peak of the S^+ ions was more dominant than the forward peak at $\phi = 0$. This behavior is reversed by changing ϕ from 0 to π. A clear periodicity of 2π with considerably large contrast was observed in the I_f/I_b ratio for OC^+ and for S^+. The phase dependencies between the OC^+ and the S^+ cations were completely out of phase with each other. This result demonstrates that oriented molecules were detected with discrimination of their head–tail order.[61] The selectivity of the oriented molecules reached 86% ($I_f/I_b = 5.9$) for OC^+. We performed simultaneous measurements using gas mixtures of OCS and reference CH_3Br to discriminate whether the orientation process was OSMI or DMO (the permanent dipole of OCS (CH_3Br) points from the small-amplitude part (large-amplitude part) to the large-amplitude part (small-amplitude part) of the wavefunction[61]) (Figs. 20(c) and 20(d)). The experimental result showed that there is a definite correlation between the orientation of detected molecules and the orbital asymmetry, where the S^+ in OCS and Br^+ in CH_3Br were completely in phase with each other. Moreover, the directions of the detected molecules are consistent with those expected by the molecular ADK model. Even for nanosecond pulses, which have sufficient time for DMO, OSMI is the main contributor to the orientation process.

Therefore, we have experimentally confirmed that OSMI induced by directionally asymmetric tunneling ionization is free from laser wavelength constraint and is observed universally in a vast range of pulse durations in the femtosecond–nanosecond regime. Additionally, many other studies concerning the interaction between molecules and intense nanosecond laser fields have confirmed that molecules can be dynamically aligned (while not discriminating the head–tail order of molecules) through the interaction between nonresonant laser fields and induced dipoles.[89] Therefore, it is reasonable to expect that an intense nanosecond $\omega + 2\omega$ laser field can induce OSMI in dynamically aligned molecules, rather than in randomly oriented molecules, during the laser pulse.[61]

6. Summary

We have investigated the interaction between gas-phase molecules with asymmetric structure and intense (10^{12-13} W/cm^2) phase-controlled $\omega + 2\omega$ pulses with an asymmetric waveform. We observed OSMI, which is impossible to achieve with a monochromatic laser field with a symmetric waveform. The direction of oriented molecules can be easily flipped by changing the relative phase difference $(0, \pi)$. We have experimentally demonstrated that, as a consequence of directionally asymmetric TI, OSMI induced by phase-controlled $\omega + 2\omega$ laser fields reflects the asymmetric geometry of the HOMO structure. The present experiments were performed under the condition of Keldysh parameter $\gamma \sim 2$, which can be categorized as an intermediate region between the TI region and the MPI region. Although molecular ADK theory is quantitatively valid only in the region of $\gamma < 1$, the theory seems to be applicable for quantitative discussions on OSMI in the present study. OSMI can be achieved through discrimination of the wavefunction in the space domain by the enhancement of nonlinear interaction between the asymmetric laser fields and the asymmetric HOMO structure. Notably, OSMI is free of laser wavelength constraints and is observed over a wide range of pulse durations in the femtosecond–nanosecond regime. Furthermore, OSMI is free of the constraints of size, weight, and polarity of molecules, and this is an advantage compared to DMO, with which it is difficult to orient large, heavy molecules that require large torques at practical laser intensities, and with which it is impossible to orient nonpolar molecules with asymmetric structures. Moreover, the directionally asymmetric TI can manipulate the directionality of photoelectrons and ionization time in the attosecond time region. This method provides a powerful tool for tracking the quantum dynamics of photoelectrons by using phase-dependent oriented molecules as a phase reference in simultaneous ion–electron detection.

Acknowledgments

The author thanks M. Tachiya, T. Nakanaga, F. Ito, N. Saito, H. Nonaka, S. Ichimura, and Toru Morishita. This work was supported by the Fund for Young Researchers from the Ministry of Education, Culture, Sports, Science and Technology (MEXT); the Mitsubishi Foundation; the Sumitomo Foundation; the Precursory Research for Embryonic Science and Technology (PRESTO) program from Japan Science and Technology (JST); and a Grant-in-Aid for Young Scientists (A), Young Scientists (B), Challenging Exploratory Research, and Scientific Research (B) from the Japan Society for the Promotion of Science (JSPS).

References

1. W. T. Hill and C. H. Lee, *Light-Matter Interaction: Atoms and Molecules in External Fields and Nonlinear Optics* (WILEY-VCH, Weinheim, 2007), ISBN: 978-3-827-40661-6.
2. F. H. Faisal, *Theory of Multiphoton Processes* (Plenum, New York, 1987), ISBN: 0-306-42317.
3. W. R. Zipfel, R. M. Williams and W. W. Webb, *Nat. Biotechnol.* **21**, 1369 (2003) and references therein.
4. L. V. Keldysh, *Sov. Phys. JETP* **20**, 1307 (1965).
5. A. M. Perelomov, V. S. Popov and M. V. Terent'ev, *Sov. Phys. JETP* **23**, 924 (1966).
6. M. V. Ammosov, N. B. Delone and V. P. Krainov, *Sov. Phys. JETP* **64**, 1191 (1987).
7. P. B. Corkum, N. H. Burnett and F. Brunel, *Phys. Rev. Lett.* **62**, 1259 (1989).
8. E. Mevel, P. Breger, R. Trainham, G. Petite and P. Agostini, *Phys. Rev. Lett.* **70**, 406 (1993).
9. G. L. Yudin and M. Y. Ivanov, *Phys. Rev. A* **64**, 013409 (2001).
10. A. de Bohan, B, Piraux, L. Ponce, R. Taeb, V. Veniard and A. Maquet, *Phys. Rev. Lett.* **89**, 113002 (2002).
11. X. M. Tong, Z. X. Zhao and C. D. Lin, *Phys. Rev. A* **66**, 033402 (2002).
12. C. D. Lin and X. M. Tong, *J. Photochem. Photobio. A* **182**, 213 (2006).
13. A. S. Alnaser, S. Voss, X.-M. Tong, C. M. Maharjan, P. Ranitovic, B. Ulrich, T. Osipov, B. Shan, Z. Chang and C. L. Cocke, *Phys. Rev. Lett.* **93**, 113003 (2004).
14. A. S. Alnaser, C. M. Maharjan, X. M. Tong, B. Ulrich, P. Ranitovic, B. Shan, Z. Chang, C. D. Lin, C. L. Cocke and I. V. Litvinyuk, *Phys. Rev. A* **71**, 031403(R) (2005).
15. D. Pavičić, K. F. Lee, D. M. Rayner, P. B. Corkum and D. M. Villeneuve, *Phys. Rev. Lett.* **98**, 243001 (2007).
16. S.-F. Zhao, C. Jin, A.-T. Le, T. F. Jiang and C. D. Lin, *Phys. Rev. A* **80**, 051402 (2009).
17. S. Petretti, Y. V. Vanne, A. Saenz, A. Castro and P. Decleva, *Phys. Rev. Lett.* **104**, 223001 (2010).
18. R. Murray, M. Spanner, S. Patchkovskii and M. Y. Ivanov, *Phys. Rev. Lett.* **106**, 173001 (2011).
19. T. Brabec and F. Krausz, *Rev. Mod. Phys.* **72**, 545 (2000) and references therein.
20. P. B. Corkum and F. Krausz, *Nature Phys.* **3**, 381 (2007) and reference therein.
21. F. Krausz and M. Ivanov, *Rev. Mod. Phys.* **81**, 163 (2009) and references therein.
22. M. Shapiro and P. Brumer, *Principles of the Quantum Control of Molecular Processes* (John Wiley, New York, 2003).
23. M. Dantus and V. V. Lozovoy, *Chem. Rev.* **104**, 1813 (2004).
24. Y. Y. Yin, C. Chen, D. S. Elliott and A. V. Smith, *Phys. Rev. Lett.* **69**, 2353 (1992).
25. Zheng-Min Wang and D. S. Elliott, *Phys. Rev. Lett.* **87**, 173001 (2001).
26. R. Yamazaki and D. S. Elliott, *Phys. Rev. Lett.* **98**, 053001 (2007).
27. R. Yamazaki and D. S. Elliott, *Phys. Rev. A* **76**, 053401 (2007).
28. Y. Y. Yin, D. S. Elliott, R. Shehadeh and E. R. Grant, *Chem. Phys. Lett.* **241**, 591 (1995).
29. G. Kurizki, M. Shapiro and P. Brumer, *Phys. Rev. B* **39**, 3435 (1989).
30. E. Dupont, P. B. Corkum, H. C. Liu, M. Buchanan and Z. R. Wasilewski, *Phys. Rev. Lett.* **74**, 3596 (1995).

31. A. Hache, Y. Kostoulas, R. Atanasov, J. L. P. Hughes, J. E. Sipe and H. M. van Driel, *Phys. Rev. Lett.* **78**, 306 (1997).
32. N. L. Manakov, V. D. Ovsiannikov and A. F. Starace, *Phys. Rev. Lett.* **82**, 4791 (1999).
33. M. Gunawardena and D. S. Elliott, *Phys. Rev. Lett.* **98**, 043001(2007).
34. M. Gunawardena and D. S. Elliott, *Phys. Rev. A* **76**, 033412 (2007).
35. A. Bolovinos, S. Cohen and I. Liontos, *Phys. Rev A* **77**, 023413 (2008).
36. E. Charron, A. Giusti-Suzor and F. H. Mies, *Phys. Rev. Lett.* **75**, 2815 (1995).
37. E. Charron, A. Giusti-Suzor and F. H. Mies, *J. Chem. Phys.* **103**, 7359 (1995).
38. B. Sheehy, B. Walker and L. F. DiMauro, *Phys. Rev. Lett.* **74**, 4799 (1995).
39. K. J. Schafer and K. Kulander, *Phys. Rev. A* **45**, 8026 (1992).
40. N. B. Baranova, H. R. Reiss and B. Ya. Zel'dovich, *Phys. Rev. A* **48**, 1497 (1993).
41. D. W. Schumacher, F. Weihe, H. G. Muller and P. H. Bucksbaum, *Phys. Rev. Lett.* **73**, 1344 (1994).
42. D. W. Schumacher and P. H. Bucksbaum, *Phys. Rev. A* **54**, 4271 (1996).
43. D. Ray, Z. Chen, S. De, W. Cao, I. V. Litvinyuk, A. T. Le, C. D. Lin, M. F. Kling and C. L. Cocke, *Phys. Rev. A* **83**, 013410 (2011).
44. M. R. Thompson, M. K. Thomas, P. F Taday, J. H. Posthumus, A. J. Langley, F. J. Frasinski and K. Codling, *J. Phys. B* **30**, 5755 (1997).
45. A. D. Bandrauk and S. Chelkowski, *Phys. Rev. Lett.* **84**, 3562 (2000).
46. S. Chelkowski, M. Zamojski and A. D. Bandrauk, *Phys. Rev. A* **63**, 023409 (2001).
47. D. Ray, F. He. S. De, W. Cao, H. Mashiko, P. Ranitovic, K. P. Singh, I. Znakovskaya, U. Thumm, G. G. Paulus, M. F. Kling, I. V. Litvinyuk and C. L. Cocke, *Phys. Rev. Lett.* **103**, 223201 (2009).
48. K. J. Betsch, D. W. Pinkham and R. R. Jones, *Phys. Rev. Lett.* **105**, 223002 (2010).
49. H. Li, D. Ray, S. De, I. Znakovskaya, W. Cao, G. Laurent, Z. Wang, M. F. Kling, A. T. Le and C. L. Cocke, *Phys. Rev. A* **84**, 043429 (2011).
50. M. J. J. Vrakking and S. Stolte, *Chem. Phys. Lett.* **271**, 209 (1997).
51. C. M. Dion, A. D. Bandrauk, O. Atabek, A. Keller, H. Umeda and Y. Fujimura, *Chem. Phys. Lett.* **302**, 215 (1999).
52. T. Kanai and H. Sakai, *J. Chem. Phys.* **115**, 5492 (2001).
53. S. Guérin, L. P. Yatsenko, H. R. Jauslin, O. Faucher and B. Lavorel, *Phys. Rev. Lett.* **88**, 233601 (2002).
54. Tehini and Sugny, *Phys. Rev. A* **77**, 023407 (2008).
55. S. De, I. Znakovskaya, D. Ray, F. Anis, Nora G. Johnson, I. A. Bocharova, M. Magrakvelidze, B. D. Esry, C. J. Cocke, I. V. Litvinyuk and M. F. Kling, *Phys. Rev. Lett.* **103**, 153002 (2009).
56. K. Oda, M. Hita, S. Minemoto and H. Sakai, *Phys. Rev. Lett.* **104**, 213901 (2010).
57. H. Ohmura, T. Nakanaga and M. Tachiya, *Phys. Rev. Lett.* **92**, 113002 (2004).
58. H. Ohmura and T. Nakanaga, *J. Chem. Phys.* **120**, 5176 (2004).
59. H. Ohmura, N. Saito and M. Tachiya, *Phys. Rev. Lett.* **96**, 173001 (2006).
60. H. Ohmura, F. Ito and M. Tachiya, *Phys. Rev. A* **74**, 043410 (2006).
61. H. Ohmura and M. Tachiya, *Phys. Rev. A* **77**, 023408 (2008).
62. H. Ohmura, N. Saito, H. Nonaka and S. Ichimura, *Phys. Rev. A* **77**, 053405 (2008).
63. H. Ohmura, N. Saito and T. Morishita, *Phys. Rev. A* **83**, 063407 (2011).
64. H. Ohmura, Directionally asymmetric tunneling ionization and control of molecular orientation by phase-controlled laser fields, Chapter 5, In *Progress in Ultrafast Intense*

Laser Science VII, pp. 109–126, ISBN: 978-3-642-18326-3.

65. P. Agostini, F. Fabre, G. Mainfray, G. Petite and N. Rahman, *Phys. Rev. Lett.* **42**, 1127 (1979).

66. R. R. Freeman, T. J. McIlrath, P. H. Bucksbaum and M. Bashkansky, *Phys. Rev. Lett.* **57**, 3156 (1986).

67. J. R. Oppenheimer, *Phys. Rev.* **31**, 66 (1928).

68. L. D. Landau and E. M. Lifshitz, Kvantovaya mekhanika (Quantum Mechanics), Fizmatgiz, 1963.

69. B. M. Smirnov and M. I. Chibisov, *Sov. Phys. JETP* **22**, 585 (1966).

70. P. W. Milonni and J. R. Ackerhalt, *Phys. Rev. A* **39**, 1139 (1989).

71. F. Trombetta, S. Basile and G. Ferrante, *Phys. Rev. A* **40**, 2774 (1989).

72. F. H. M. Faisal, *J. Phys. B* **6**, L89 (1973).

73. H. R. Reiss, *Phys. Rev. A* **22**, 1786 (1980).

74. M. Uiberacker, Th. Uphues, M. Schultze, A. J. Verhoef, V. Yakovlev, M. F. Kling, J. Raushenberger, N. M. Kabachnik, H. Schröder, M. Lezius, K. L. Kompa, H.-G. Muller, M. J. J. Vrakking, S. Hendel, U. Kleineberg, U. Heinzmann, M. Drescher and F. Krausz, *Nature* **446**, 627 (2007).

75. P. Eckle, A. N. Pfeiffer, C. Cirelli, A. Staudte, R. Dörner, H. G. Muller, M. Büttiker and U. Keller, *Science* **322**, 1525 (2008).

76. Gaussian 03, Revision C.02, M. J. Frisch, G. W. Trucks, H. B. Schlegel, G. E. Scuseria, M. A. Robb, J. R. Cheeseman, J. A. Montgomery, Jr., T. Vreven, K. N. Kudin, J. C. Burant, J. M. Millam, S. S. Iyengar, J. Tomasi, V. Barone, B. Mennucci, M. Cossi, G. Scalmani, N. Rega, G. A. Petersson, H. Nakatsuji, M. Hada, M. Ehara, K. Toyota, R. Fukuda, J. Hasegawa, M. Ishida, T. Nakajima, Y. Honda, O. Kitao, H. Nakai, M. Klene, X. Li, J. E. Knox, H. P. Hratchian, J. B. Cross, C. Adamo, J. Jaramillo, R. Gomperts, R. E. Stratmann, O. Yazyev, A. J. Austin, R. Cammi, C. Pomelli, J. W. Ochterski, P. Y. Ayala, K. Morokuma, G. A. Voth, P. Salvador, J. J. Dannenberg, V. G. Zakrzewski, S. Dapprich, A. D. Daniels, M. C. Strain, O. Farkas, D. K. Malick, A. D. Rabuck, K. Raghavachari, J. B. Foresman, J. V. Ortiz, Q. Cui, A. G. Baboul, S. Clifford, J. Cioslowski, B. B. Stefanov, G. Liu, A. Liashenko, P. Piskorz, I. Komaromi, R. L. Martin, D. J. Fox, T. Keith, M. A. Al-Laham, C. Y. Peng, A. Nanayakkara, M. Challacombe, P. M. W. Gill, B. Johnson, W. Chen, M. W. Wong, C. Gonzalez and J. A. Pople, Gaussian, Inc., Wallingford CT, 2004.

77. L. Holmegaard, J. L. Hansen, L. Kalhøj, S. L. Kragh, H. Stapelfeldt, F. Filsinger, J. Küpper, G. Meijer, D. Dimitrovski, M. Abu-samha, C. P. J. Martiny and L. B. Madsen, *Nature Phys.* **6**, 428 (2010).

78. D. Dimitrovski, C. P. J. Martiny and L. B. Madsen, *Phys. Rev. A* **82**, 053404 (2010).

79. M. Abu-samha and L. B. Madsen, *Phys. Rev. A* **82**, 043413 (2010).

80. D. Dimitrovski, M. Abu-samha, L. B. Madsen, F. Filsinger, G. Meijer, J. Küpper, L. Holmegaard, L. Kalhøj, J. H. Nielsen and H. Stapelfeldt, *Phys. Rev. A* **83**, 023405 (2011).

81. J. Ullrich, R. Moshammer, A. Dorn, R. Dörner, L. Ph. H. Schmidt and H. Schmidt-Böcking, *Rep. Prog. Phys.* **66**, 1463 (2003).

82. J. Lavancier, D. Normand, C. Cornaggia, J. Morellec and H. X. Liu, *Phys. Rev. A* **43**, 1461 (1991).

83. J. Wu, L. Ph. H. Schmidt, M. Kunitski, M. Mecke, S. Voss, H. Sann, H. Kim, T. Jahnke, A. Czasch and R. Dörner, *Phys. Rev. Lett.* **108**, 183001 (2012).
84. M. J. DeWitt and R. J. Levis, *Phys. Rev. Lett.* **81**, 5101 (1998).
85. H. R. Reiss, *Phys. Rev. A* **75**, 031404 (2007).
86. B. K. McFarland, J. P. Farrell, P. H. Bucksbaum and M. Gühr, *Science* **322**, 1232 (2008).
87. H. Akagi, T. Otobe, A. Staudte, A. Shiner, F. Turner, R. Dörner, D.M. Villeneuve and P. B. Corkum, *Science* **325**, 1364 (2009).
88. I. Znakovskaya, P. von den Hoff, S. Zherebtsov, A. Wirth, O. Herrwerth, M. J. J. Vrakking, R. de Vivie-Riedle and M. F. Kling, *Phys. Rev. Lett.* **103**, 103002 (2009).
89. T. Morishita, Z. Chen, S. Watanabe and C. D. Lin, *Phys. Rev. A* **75**, 023407 (2007).
90. H. Stapelfeldt and T. Seideman, *Rev. Mod. Phys.* **75**, 543 (2003) and references therein.

CHAPTER 3

REACTION AND IONIZATION OF POLYATOMIC MOLECULES INDUCED BY INTENSE LASER PULSES

D. Ding[*], C. Wang[*], D. Zhang[*], Q. Wang[*,†],
D. Wu[*] and S. Luo[*]

Interaction of atoms or molecules with intense laser fields is an emerging subject of atomic, molecular, and optical physics. Investigation in dynamics of these systems will be able to uncover various new phenomena and change our way of controlling the evolution of matter in microscale. This chapter summarizes a number of the studies in recent years and is intended for authors to explain some basic features of polyatomic molecules in intense laser fields and their dynamic processes induced by femtosecond laser pulses. It is hoped that this chapter is informative and gives the readers some insight into this field of fundamental science.

1.1. Introduction

Since its invention in the early 1960s, laser has been developed as a tool for scientific studies and technical applications due to the reason that it can deliver energy in controllable ways. Many technique breakthroughs, such as Q-switching, mode-locking, and chirped pulse amplification (CPA), enable the laser pulse to become shorter, giving a dramatic increase of laser intensity (pulse powers per unit area, W/cm^2). Among them, CPA emerged in the mid-1980s as a solution for overcoming the limitation from the laser amplifier operating at high intensity. High laser flux may induce significant self-focusing and self-phase modulation, resulting in optical

[*]Institute of Atomic and Molecular Physics, Jilin University, Changchun 130012, China
[†]State Key Laboratory of Theoretical Physics, Institute of Theoretical Physics, CAS, Beijing 100190, China

damage of the laser. CPA is realized by manipulating the ultrashort pulse in a controllable and reversible fashion, i.e., stretching, amplifying, and subsequently recompressing, so that the laser amplifier never encounters a short, high power pulse, and only the laser system components compatible with such high peak powers can be exposed to it (see, for example, the review paper of Mourou *et al.*).[1] This technique is remarkable and revolutionary, and nowadays a tabtop femtosecond laser in laboratory can deliver an output power up to 10^{18} W/cm^2. Also pulses as short as a few femtoseconds (fs, 10^{-15} s), so-called few-cycle pulses, have been directly generated by a Ti:sapphire laser controlling hollow fiber compressor.[2] All these achievements in laser technology have opened a new domain of physics and chemistry for exciting, probing, and controlling matter and its dynamics in a precision of atomic scale.

As an electromagnetic wave, optical electric fields associated with the peak powers of ultrashort laser pulses are extremely high. From Maxwell's equation, the relation of the peak electric field strength, E (V/cm), with the laser intensity, I (W/cm^2), is formulated by

$$E \approx 27.4 I^{1/2}. \qquad (1.1)$$

The electric field strength E brought by intense fs-laser pulses is comparable to or even exceeds the Coulombic binding fields inside atoms and molecules. Considering its electron at the orbital of the ground state in atomic hydrogen, the strength of interacting Coulombic field is $E_a = m^2 e^5/\hbar^4 = 5.14 \times 10^9$ V/cm (the atomic unit of electric field intensity), which corresponds to a laser intensity of 3.51×10^{16} W/cm^2, at accessible levels simply by a compact fs-laser system. Taking molecular bond energy of one order less than its electron binding energy into account, comparable field strength is easily delivered by the laser system.

Therefore, in such intense laser fields, laser-matter interaction is non-perturbative, nonlinear, and even relativistic with the increase in laser intensity (Figure 1). In high nonlinear interaction region, atoms or molecules can absorb multiple photons simultaneously to be ionized (multiphoton ionization, MPI, or above-threshold ionization, ATI), or the electrons can also be released simply by barrier-suppression in intense fields (field ionization or tunneling ionization).[3,4] According to the nature of interaction, intense laser fields can be simply classified by a parameter

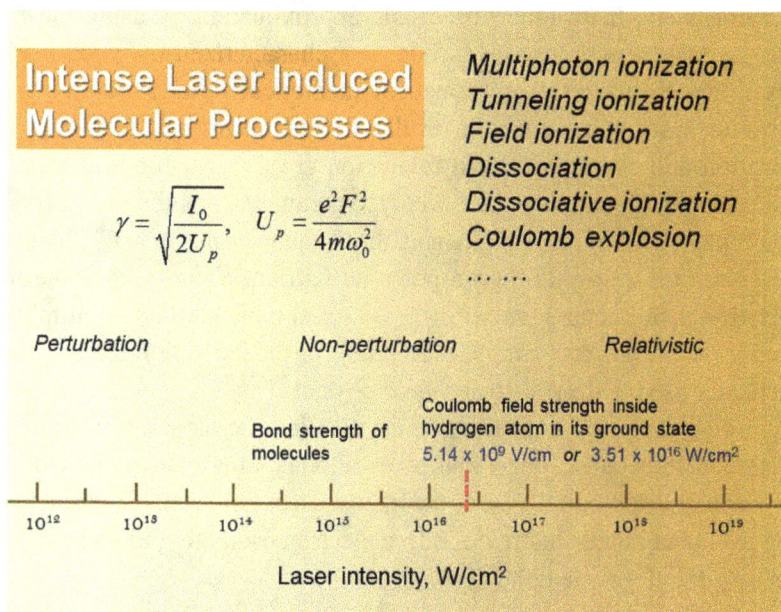

Fig. 1.   Intense laser induced molecular processes in which the interaction can be treated as perturbative, nonperturbative or even relativistic, depending on the laser intensity interacted with molecules.

γ, defined by Keldysh,[5]

$$\gamma = \frac{\omega}{eE}\sqrt{2mI_0} = \frac{1}{2K_0 F},\qquad (1.2)$$

where I_0 is the ionization potential of atoms, E and ω are the amplitude (or field strength) and the frequency of the electric wave field $E(t) = E\cos\omega t$, respectively, F is the reduced field strength $F = E/\kappa^3 E_a$ with $\kappa = (I_0/I_H)^{1/2}$, $I_H = me^4/2\hbar^2 = 13.6\,\mathrm{eV}$ the ionization potential of atomic hydrogen, and $K_0 = I_0/\hbar\omega$ is the minimal number of photons required for ionization. Keldysh assumed the total wavefunction as a sum of the wavefunctions for the ground state and the Volkov continuum (in which a harmonic move of the released electron with time in the linearly polarized electric field of the optical pulses is included while the Coulomb interaction between the ejected electron and the atomic core is neglected[6]) and gave analytically the direct photoionization rate for atoms in a strong electromagnetic field under the dipole approximation by using the first-order perturbation theory. This rate was characterized by γ parameter. This is the first time of systematical theoretical description of atomic ionization in strong field and the results showed that the field or "tunneling" ionization and the multiphoton ionization are the two limiting cases of nonlinear photoionization since Keldysh parameter γ is the ratio between the frequency ω of laser light and the frequency $\omega_t = eE/(2mI_0)^{1/2}$ of electron tunneling through a potential barrier. When $\gamma \ll 1$ the field ionization is dominated while for $\gamma \gg 1$, the ionization is a multiphoton process. This leads to a simple estimation for the ionization feature and is practically used widely in strong-field physics.

In discussing the ionization along with intense laser interaction with atoms and molecules, one often takes another important parameter, the ponderomotive potential, U_p, of the intense laser field which is equal to the time-averaged kinetic energy of a free electron oscillating in an ac field of intensity I and wavelength λ, i.e.,

$$U_p = \frac{e^2 F^2}{4m\omega^2} = 9.33 \times 10^{-14} I\lambda^2 \, [eV]. \tag{1.3}$$

By using ponderomotive potential U_p, the Keldysh parameter γ is also expressed as a ratio of the applied field to the ionization potential, $\gamma = \omega/\omega_t = \sqrt{I_0/2U_p}$. Therefore, it is obvious that one can use U_p to classify the laser intensity interacted with atoms or molecules.

Tunneling ionization is a quantum phenomenon, forbidden by classic laws. In tunneling ionization process, electrons in an atom or molecule can pass through a potential barrier and escape with a certain probability even when they do not have sufficient energy over the barrier. This tunneling process occurs when the atomic or molecular Coulombic potential barrier is distorted and its length along which the electrons have to pass decreases by applied intense laser electric field. Though multiphoton ionization was observed long time ago, the tunneling ionization of atoms was observed first by Chin et al.[7] in rare-gas atoms. Based on tunneling ionization, Corkum[8] proposed a three-step or rescattering model for interpreting various phenomena of atoms in strong fields. In this simple picture, the atom is ionized first by the laser field to produce a free electron and a residual ion, then the electron is accelerated by the oscillating laser field and driven back to the parent ion by the field when changing its direction, and finally the electron is "rescattered" by the ion elastically, inelastically or recombined with the ion.

While many studies have exclusively being done for atomic ionization in intense laser fields, the equivalent studies on molecules are less developed. Though many molecular phenomena is parallel to the atomic cases, the situation is much more complex to model theoretically and observe experimentally for molecular processes in intense laser fields. Molecules contain additional nuclear degrees of freedom and, consequently, nuclear rotational and vibrational dynamics need to be taken into account. For example, in evaluating the Keldysh parameters of molecules in intense laser fields, the influence of molecular electronic orbital shape, size, and polarization should be considered,[9a,9b] this leads to an increase of field ionization probability in the case of polyatomic molecules compared with that of atoms. Furthermore, ionization of atoms has been described by single-active electron (SAE), see reference by Schafer *et al.*,[10] and strong-field approximation (SFA), see reference by Lewenstein *et al.*[11] very successfully, but the theories for dealing with many phenomena of molecular ionization are still inadequate in the case of intense laser fields and more general theoretical approaches are required to interpret or model various new experimental observations. Studies on molecular processes in intense femtosecond laser fields will help to understand the physics behind many molecular processes observed, for example, multi-electron effect, coupling of electronic-vibrational movement, stereo effects on molecular ionization/dissociation, etc.

In this chapter, we focus on "moderate" intense laser induced processes of molecular systems, i.e., in the laser intensity region of $10^{12} \sim 10^{14}$ W/cm^2, with $40 \sim 100$ fs pulse duration, in which most of the optical field induced molecular processes are covered and many of them are still unclear. For the correlated many-body system of molecules interacted with intense laser fields, it is a very challenging task to describe theoretically and measure completely. Even though mainly summarized ideas are based on measurement of ions, the effect of nonspherical symmetry of molecular electronic orbitals is noticeable and the differences from respective atomic like theories are generally found. For photoelectron measurement which is beyond the scope of our present consideration, readers are referred to the reviews of Stolow *et al.*,[12] Wollenhaupt *et al.*,[13] and the recent one by Krausz and Ivanov.[14]

1.2. Ionization Rate of Molecules in Intense Laser Fields

Many studies have been carried out during past decades to improve our pictures of laser-matter interaction through quantitative comparisons between experiment and theory. Understanding the ionization rates plays an important role in exploring the pictures of mechanics and developing the theoretical description for interaction processes of atoms or molecules with intense laser fields.[3,4,15,16] One example is that suppressed molecular ionization was found for D_2 and O_2 in comparison of rate constants with their companion atoms Ar and Xe,[17] illustrating the emphasis on molecular orbital features and multi-center interference is necessary. This section will summarize briefly the theoretical development and then give some comparative studies on ionization rates for polyatomic molecules in fs-laser fields.

1.2.1. *Theoretical approaches for ionization rates of molecules in intense laser fields*

Theoretical methods have been developed for calculating the ionization rates of atoms or molecules in intense laser fields for many years. The quantum theoretical calculations of the ionization rates for atoms are readily available now-a-days. But, considering molecules with the additional degrees of freedom in nuclear vibration and rotation, and nonspherically symmetric electron orbitals, *ab initio* calculations for the ionization rates in intense laser fields are extremely difficult in solving time-dependent Schrödinger equation or very computationally demanding in numerical procedures, and therefore, theoretical approaches need to be developed with some approximations. In general, two of most common assumptions for the case of atoms are single-active electron (SAE) and strong-field approximation (SFA). In the SAE approximation only a single electron is considered to move in the potential created by the nucleus and the remaining electrons of the atom which are frozen in their ground state orbital.[18,19] In the SFA, it is assumed that the electronic continuum wavefunction is coupled much strongly to the field than to the residual ion so that the Coulomb field can be neglected after ionization.[20,21]

On the other hand, various approaches such as Keldysh,[5] PPT theory,[22] KFR,[21] and ADK[23] are established for the case of atoms for years. In general, one can take these atomic-like models in the first-order approximation and compare the ionization rates of molecules with respect to these atoms that have nearly identical binding energies since the ionization rate depends critically on the ionization potential of the atom or molecule. Because SAE and SFA play their role for molecules in intense laser fields,[24] as in the case of atoms, several theoretical methods are also developed based on the SAE approximation, following the tunneling theory (MO-ADK[16]) or KFR theory with SFA (g-KFR,[25,26,27]) in which Born–Oppenheimei approximation and molecular orbital are also adapted. These calculations are widely applied for interpreting and/or comparing the observations although in the further calculations it is necessary to include the effects involved with vibronic movement and interference from many atom-centers of molecules. We have outlined it here only to put these approaches of the theories in context and a more complete and detailed review may be found in the references cited here, the book of Grossmann[28] and other related chapters in the books of this series.

Hamiltonian for the atoms in laser fields is described by

$$\hat{H}_F(\vec{r}, t) = -\frac{\hbar^2 \nabla^2}{2m} + V(\vec{r}) - \vec{d} \cdot \vec{F}(t) \tag{1.4}$$

where $\vec{d}$ is the dipole moment, $\vec{F}(t)$ is electric field given by

$$\vec{F}(t) = \vec{F} \cos \omega t. \tag{1.5}$$

Thus, the time-dependent Schrödinger equation is written as

$$i\hbar \frac{\partial}{\partial t} \Psi(\vec{r}, t) = H_F(\vec{r}, t) \Psi(\vec{r}, t). \tag{1.6}$$

Omitting the potential term in the Hamiltonian, its solution should be in the form of,

$$\psi_{\vec{p}}(\vec{r}, t) = \exp\left[\frac{i}{\hbar}\left\{[\vec{p} - e\vec{A}(t)] \cdot \vec{r} - \frac{1}{2m}\int_0^t dt'[\vec{p} - e\vec{A}(t')]^2\right\}\right] \tag{1.7}$$

This is the Volkov function with $\vec{p}$ the momentum of the released electrons.[6]

According to Keldysh,[5] the total wavefunction is a sum of the wavefunctions of the ground-state and of free electrons, i.e.,

$$\Psi(\vec{r}, t) = \psi_g(\vec{r}) \exp\left(-\frac{i}{\hbar} E_g t\right) + \int d^3 p c_{\vec{p}}(t) \psi_{\vec{p}}(\vec{r}, t), \qquad (1.8)$$

$$c_{\vec{p}}(t) = \frac{i}{\hbar} \int_0^t dt' \langle \psi_{\vec{p}}(\vec{r}, t') | \vec{d} \cdot \vec{F} | \psi_g(\vec{r}) \rangle \exp\left(-\frac{i}{\hbar} E_g t'\right) \cos \omega t', \qquad (1.9)$$

where $\psi_g(\vec{r})$ is the ground-state wavefunction, E_g is the eigenenergy of ground state, $\vec{F}$ is the amplitude of the incoming optical wave. When the ground-state is assigned as the $1s$ state for a hydrogen-like atom,

$$\psi_g(\vec{r}) = \sqrt{\frac{1}{\pi a^3}} \exp\left(-\frac{r}{a}\right), \quad a = \frac{a_0}{Z}, \qquad (1.10)$$

the Keldysh theory gives the photoionization rate of the ground-state hydrogen atom in the dipole approximation as

$$k(\vec{F}) = \frac{2}{\hbar^2} \lim_{T \to \infty} \mathrm{Re} \int \frac{d^3 p}{(2\pi\hbar)^3} \int_0^T dt \cos(\omega t) \cos(\omega T)$$

$$\times V_0^* \left(\vec{p} + \frac{e\vec{F}}{\omega} \sin \omega T\right) V_0 \left(\vec{p} + \frac{e\vec{F}}{\omega} \sin \omega t\right)$$

$$\times \exp\left[\frac{i}{\hbar} \int_T^t d\tau \left\{ I_0 + \frac{1}{2m} \left(\vec{p} + \frac{e\vec{F}}{\omega} \sin \omega \tau\right)^2 \right\} \right] \qquad (1.11)$$

where the ionization potential I_0 is the $1s$ state energy of the hydrogen-like atom,

$$I_0 = -E_g = \frac{Z^2 e^2}{2a_0} \qquad (1.12)$$

and

$$V_0(\vec{p}) = 8i(\pi a^3)^{1/2} e\hbar \vec{F} \cdot \vec{\nabla}_{\vec{p}} \left(1 + \frac{p^2 a^2}{\hbar^2}\right)^{-2}. \qquad (1.13)$$

Furthermore by using residue theorem, the formula of the atomic photoionization rate can be given as:

$$k = 4\sqrt{\frac{2I_0\omega}{\hbar}} \left(\frac{\gamma}{\sqrt{1+\gamma^2}}\right)^{3/2} N(\gamma, \omega, I_0, \tilde{I}_0, B, C)$$

$$\times \exp\left[-\frac{2\tilde{I}_0}{\hbar\omega}\left(\sinh^{-1}\gamma - \frac{\gamma\sqrt{1+\gamma^2}}{1+2\gamma^2}\right)\right], \qquad (1.14)$$

where $N(\gamma, \omega, I_0, \tilde{I}_0, B, C)$ is a pre-exponential factor and

$$\tilde{I}_0 = I_0 + U_p, \quad U_p = \frac{e^2 F^2}{4m\omega^2}. \qquad (1.15)$$

Faisal[29] and Reiss,[21,30] established the formula for calculating the rates from the velocity gauge Keldysh theory. This KFR theory is simple and practical in calculating the rates than its original Keldysh theory, and therefore to be used frequently. Similarly, the ionization continuum is treated by Volkov function and the general Bessel function J_N is employed for calculating the integrals in KFR theory, then the ionization rates is formulated as

$$k_H(\omega) = \int \frac{d^3p}{(2\pi\hbar)^3} \hat{\chi}_{1s}^*(\vec{p}, a_0) \hat{\chi}_{1s}(\vec{p}, a_0) \left(\frac{p^2}{2m} + I_H\right)$$

$$\times \sum_{N=-\infty}^{\infty} \left\{ \frac{J_{N+n}J_N^*}{p^2/(2m) + \tilde{I}_H - N\hbar\omega + i\varepsilon} \right.$$

$$\left. - \frac{J_{N-n}J_N^*}{p^2/(2m) + \tilde{I}_H - N\hbar\omega + i\varepsilon} \right\} \qquad (1.16)$$

In Keldysh or KFR theory, the formulas are established based on a hydrogen-like atomic model and thus, only atomic ionization can be calculated. For molecular ionization, the case is more complex. Lin and his collaborators generalized the KFR theory and applied it to the case of molecular ionization by combining the theory with molecular orbital (MO) theory and Born–Oppenheimer approximation.[25,27] Thus, this g-KFR theory can take into account many electron features (i.e., to reduce to the one-electron problem) by molecular orbital method and treat the effect of nuclear motion (through the vibrational overlap integral) by Born–Oppenheimer approximation.

In the g-KFR approach it is assumed that the ground electronic state of molecule or molecular cation is well described in terms of molecular orbitals obtained from *ab initio* calculation and for the ionized state the electron wave function is described by the Volkov continuum state. Then, the total electronic wave function of the molecule or molecular cation is expressed as

$$\Psi_M(r, R, t) = \psi_g(r, R) \exp\left(-\frac{i}{\hbar} E_g t\right)$$
$$+ \int \frac{d^3 p}{(2\pi\hbar)^3} c_{\vec{p}}(t) \psi_{\vec{p}}(r, R, t) \exp\left(-\frac{i}{\hbar} E_p t\right),$$

where r refers to electronic and R to nuclear coordination, $\psi_g(r, R)$, $\psi_p(r, R, t)$ is the neutral molecular wavefunction and molecular cation wavefunction, respectively.

$$\psi_g(r, R) = \|\chi_{1s}(1)\alpha(1)\chi_{1s}(2)\beta(2) \cdots \chi_{\text{HOMO}}(N_e - 1)$$
$$\times \alpha(N_e - 1)\chi_{\text{HOMO}}(N_e)\beta(N_e)\|$$

with N_e is the number of electrons, and

$$\psi_p(r, R) = c_1\|\chi_{1s}(1)\alpha(1)\chi_{1s}(2)\beta(2) \cdots \chi_{\text{HOMO}}(N_e - 1)$$
$$\times \alpha(N_e - 1)\chi_p(N_e)\beta(N_e)\|$$
$$+ c_2\|\chi_{1s}(1)\alpha(1)\chi_{1s}(2)\beta(2) \cdots \chi_p(N_e - 1)$$
$$\times \alpha(N_e - 1)\chi_{\text{HOMO}}(N_e)\beta(N_e)\|$$

with $c_1 = c_2 = -1/\sqrt{2}$. For $i = N_e - 1, N_e$,

$$\chi_{\text{HOMO}}(i) = \sum_{j=1}^{N_n} b_{j,2p}\chi_{j,2p}(\vec{r}_i - \vec{R}_j)$$

$$\chi_{\vec{p}}(i) = \exp\left[\frac{i}{\hbar}\left(\vec{p}\cdot\vec{r}_i - \frac{1}{2m}\int_{-\infty}^{t} dt'(\vec{p} - e\vec{A}(t'))^2\right)\right]. \qquad (1.17)$$

Therefore, by using similar treatment of Keldysh and KFR theories and under the assumption that the ionization only takes place from the HOMO, the photoionization rate constant can be formulated,[25,31] given as

$$
k(\vec{F}) = 2\pi S^2 \sum_{j,j'=1}^{N_e} c_j c_{j'}^* \int \frac{d^3 p}{(2\pi)^3} \hat{\chi}_j(\vec{p}) \hat{\chi}_{j'}^*(\vec{p})
$$

$$
\times \left(\frac{p^2}{2m_e} + I_0 \right)^2 \left| J_N \left(\frac{e\vec{F} \cdot \vec{p}}{m_e \omega^2}, \frac{U_p}{2\omega} \right) \right|^2
$$

$$
\times \cos\left(\vec{p} \cdot (\vec{R}_j - \vec{R}_{j'})\right) \sum_{N=-\infty}^{\infty} \delta\left(I_0 + U_p + \frac{p^2}{2m_e} - N\omega \right)
$$

$$
= \sum_N 2\pi S^2 \sum_{j,j'=1}^{N_e} c_j c_{j'}^* \int \frac{d^3 p}{(2\pi)^3} \hat{\chi}_j(\vec{p}) \hat{\chi}_{j'}^*(\vec{p})
$$

$$
\times \left(\frac{p^2}{2m_e} + I_0 \right)^2 \left| J_N \left(\frac{e\vec{F} \cdot \vec{p}}{m_e \omega^2}, \frac{U_p}{2\omega} \right) \right|^2
$$

$$
\times \cos(\vec{p} \cdot (\vec{R}_j - \vec{R}_{j'}))\delta\left(I_0 + U_p + \frac{p^2}{2m_e} - N\omega \right)
$$

$$
= \sum_N k(N) \tag{1.18}
$$

with J_N is the generalized Bessel function, c_j the coefficients of the linear combination of atomic orbitals-molecular orbital, $S = \sqrt{2}$ for the closed shell parent molecule or molecular cation, and $S = 1$ for the open shell. The g-KFR theory has been widely used to diatomic and polyatomic molecules.

On the other hand, Ammosov, Delone and Krainov[32] developed the PPT theory[22] for treating arbitrary states of hydrogen atoms in intense electromagnetic fields to the ionization rates for arbitrary atoms

$$
w = \left(\frac{3e}{\pi} \right)^{3/2} \frac{Z^2}{3n^{*3}} \frac{2l+1}{2n^* - 1} \left[\frac{4eZ^3}{(2n^* - 1)\, n^{*3} F} \right]^{2n^* - 3/2} \exp\left[\frac{-2Z^3}{3n^{*3} F} \right], \tag{1.19}
$$

with $e = 0.71828\ldots, n^*$ and l^* the effective quantum numbers. In this atomic ADK theory, the major improvement is to modify the radial wave function of the outermost electron in the asymptotic region where tunneling occurs and therefore the theory is an extension of the PPT only for hydrogen atoms to more complex atomic system. However, for a molecular system, the calculation for ionization rates is even complicated since multi-centre problem has to be treated. Based on the similar consideration on the asymptotic feature of electronic wave functions and symmetric feature,[16] Tong *et al.* expressed the molecular electronic wave functions in the asymptotic region in terms of summations of spherical harmonics in a one-center expansion,[16]

$$\psi^m(\vec{r}) = \sum_l C_l F_l(r) Y_{lm}(\hat{r}),$$

with a normalized coefficient C_l for insuring the wave function in the asymptotic region can be expressed as

$$F_l(r \to \infty) \approx r^{Z_c/\kappa - 1} e^{-\kappa r},$$

with Z_c the effective Coulomb charge, $\kappa = \sqrt{2I_p}$, and I_p the ionization potential for the given valence orbital. They realized the ADK theory calculation for the ionization rates of diatomic molecules with an arbitrary Euler angle $\vec{R}$ with respect to the low frequency ac field direction (non-aligned) is

$$w(F, R) = \left(\frac{3F}{\pi\kappa^3}\right)^{1/2} \sum_{m'} \frac{B^2(m')}{2^{|m'|}|m'|!} \frac{1}{\kappa^{2Z_c/\kappa - 1}} \left(\frac{2\kappa^3}{F}\right)^{2Z_c/\kappa - |m'| - 1} e^{-2\kappa^3/3F}$$

$$(1.20)$$

where, if $D^l_{m',m}(\vec{R})$ is the rotation matrix, one has

$$B(m') = \sum_l C_l D^l_{m',m}(\vec{R}) Q(lm'),$$

$$Q(lm) = (-1)^m \sqrt{\frac{(2l+1)(l+|m|)!}{2(l-|m|)!}}.$$

This MO-ADK method was generalized to nonlinear polyatomic molecules[33,34] previously and extended to multi-electron cases by Brabec *et al.*[35] and Zhao *et al.*[36]

1.2.2. *Experimental measurements of ionization rates of molecules and comparations with theory*

Measurement of ionization rates for molecules can be made by employing a crossed laser-molecular beam apparatus with different mass-selected charged particle detection methods such as Wiley–Mclaren type TOF mass spectrometer[31,37] or a velocity map imaging detection.[38,39] The velocity map imaging, consisting of a TOF and a 2 dimension detector, is useful since it can give the information of mass, energy, and momentum of ionic species from laser-molecule interaction (see, Figure 2).[40] With the help of multichannel, multi-hit, time-to-digital converters, photoion–photoion coincidence (PIPICO) signals can also be extracted by discriminating random coincidence contributions using momentum conservation conditions.[41]

A chirped pulse amplified Ti:sapphire laser delivers a linearly polarized laser beam with several 10 fs pulse duration and wavelength normally centered at 800 nm (frequency doubled to 400 nm). A variable attenuation for laser beam intensity can be achieved using a rotatable half-wave plate followed by a Glan–Taylor prism. Femtosecond laser pulses pick very high peak powers. Practically, the powers are concentrated further to a high level through focusing onto very small areas by using optical lenses or mirrors in laser-molecule interaction experiments. As a Gaussian beam, a laser beam of diameter D and wavelength λ gives a focal spot diameter (the beam waist) of

$$d = \frac{f\lambda}{\pi D} \sim f\theta, \tag{1.21}$$

at the focus of a lens with a focal length f in a diffraction limit. This can increase the laser intensity I by several orders of magnitude. For example, with a 350 mm focus lens, 90 fs pulse duration, 400 nm wavelength, and 1 mJ pulse energy, the beam intensity I will be a few times of 10^{14} W/cm^2 at the interaction zone.

The accurate measurement of I is critical towards to interpret and compare quantitatively experimental observations. The major problem comes from measuring the diameter d of laser beam focal spot though it can be determined simply by Eq. (1.21).

For comparison with experimental measurement, one should consider the time and spatial distribution of laser pulse as a Gaussian beam,

$$\vec{F}(t)\cos(\omega t) = \vec{F}_0 \exp(-(4\ln 2)t^2/(\Delta t_F)^2)\cos(\omega t), \qquad (1.22)$$

where Δt_F is the full width at half maximum of pulse and a Gaussian shape of the spatial distribution of the laser with the width ΔR defined by

$$I(R) = I_0 \exp(-8R^2/\Delta R^2). \qquad (1.23)$$

Then, it is clear that the ionization yield will depend on t and R, i.e., $A^+(t,R)$, and the spatial averaged ionization yield is given by

$$A^+(t) = 4\pi \int_0^\infty dR R^2 A^+(t, R). \qquad (1.24)$$

Note that here it is also assumed that, for ionization process $M \to M^+$, the rate constant k is independent of the laser time t within the pulse duration.[42]

In general, if assuming a sequential process (no electron correlation involved)

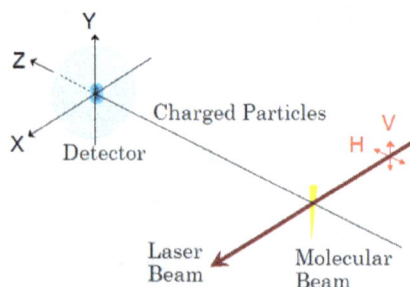

$$M \xrightarrow{k_1} M^+ \xrightarrow{k_2} M^{2+} \cdots \xrightarrow{k_n} M^{n+},$$

Fig. 2. A general experimental geometry: a linearly polarized (H-horizontal, V-vertical) laser beam intersected by a molecular beam and the produced charged particles (either ions or electrons) fly in the Z-direction and reach a detector after a certain flight time.

the photoionization rate constant k_i for the ith order parent molecular ions M^{i+} can be obtained from

$$\frac{d}{dt}A(t) = -k_1 A(t),$$

$$\cdots$$

$$\frac{d}{dt}A^{i+}(t) = k_i A^{(i-1)+}(t) - k_{i+1}A^{i+}(t),$$

$$\cdots$$

$$\frac{d}{dt}A^{n+}(t) = k_n A^{(n-1)+}(t). \tag{1.25}$$

Since the ionizing yields for the neutral and single ionized molecules are $A(t_0) = [M]_0$ and $A^{i+}(t_0) = 0$ $(i = 1, 2, \ldots, n)$, one obtains the ionization yield for the molecular ions with arbitrary ionized order by solving the rate equations of (1.25),

$$A(t) = A_0 \exp(-k_1(t - t_0)),$$

$$A^{i+}(t) = A_0 \left(\prod_{j=1}^{i} k_j\right) \sum_{j=1}^{i+1} \frac{\exp(-k_j(t - t_0))}{\prod_{l=1(\neq j)}^{i+1}(k_l - k_j)} \quad (i = 1, 2, \ldots, n-1),$$

$$A^{n+}(t) = A_0 \left(\prod_{j=1}^{n} k_j\right) \left(\frac{1}{\prod_{j=1}^{n} k_j} - \sum_{j=1}^{n} \frac{\exp(-k_j(t - t_0))}{k_j \prod_{l=1(\neq j)}^{n}(k_l - k_j)}\right).$$

$$\tag{1.26}$$

As the comparative studies, several polyatomic molecules such as cyclopentanone (C_5H_8O), methyl-substituted cyclopentanone ($C_6H_{10}O$), cyclohexanine ($C_6H_{10}O$), pyrrolidine (C_4H_9N), have been investigated.[31,37,43,44] The ionization rate constants of these molecules in intense laser fields were calculated by using various theories based on the hydrogen-like atom approach. Comparison of the different results obtained shows that the ADK and Keldysh theories overestimate the rate constants generally in the intensity range of $10^{13} \sim 10^{15}\,\text{W/cm}^2$ according to the rate limitation of ionized electron while the g-KFR theory provides a reasonable agreement with the measured relative ion yields, indicating that the g-KFR theory is useful in predicting the ionization yields of polyatomic molecules induced

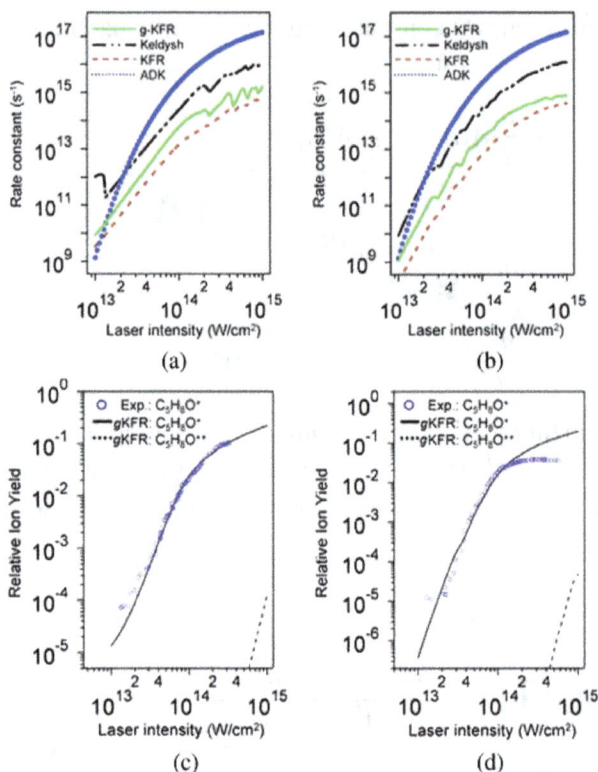

Fig. 3. Calculated results of the first ionization rate constants for cyclopentanone (C_5H_8O) by using various theories and comparison of relative ion yields between the measured (the circles) and the g-KFR calculated results (produced from the data of green curves) at $\lambda = 394$ nm (a and c) and 788 nm (b and d). In (c) and (d), the dashed lines denote the g-KFR calculated ion yields for the second ionization of the molecules (after Ref. 31).

by laser fields in the present intensity range. We take the molecular cyclopentanone (C_5H_8O) as an example for a detailed discussion as follows.

Figure 3 presents the calculated results of the various theories for first ionization rate constants and the comparison of the calculated results with the measured data for molecular cyclopentanone (C_5H_8O) irradiated by a 90 fs, 798 nm or its SHG, 394 nm, laser pulse.[31] A justification for distinction of these theories is seen in this figure clearly. The rate constants calculated by ADK (blue color) and Keldysh (black color) theories are obviously overestimated since they increase rapidly with the laser intensity and reach the values beyond $10^{16} \sim 10^{17} \, s^{-1}$ in the intensity range above $10^{14} \, W/cm^2$. This is caused by ignoring the structural characteristics of a real molecular system in these theories and can be improved when taking them into account somehow, for example, as in the case of g-KFR[25] or MO-ADK.[16] In the g-KFR theory calculation, even still in the frame of hydrogen-like model, the molecular characters have to be considered by taking the HOMO of cyclopentanone consisting mainly of $2p$ atomic orbitals and with some $3p$ orbitals since the HOMO is a nonbonding orbital of C–O with the p-orbital character. This wavefunction of the mixing (LCAO) in the g-KFR differs from that adapted in the calculation of ADK, Keldysh, or KFR. Therefore, the results obtained by the g-KFR are improved largely and a consistency of the experimental and theoretical results is achieved as shown in Figs. 3(c) and 3(d). Additionally, both the KFR and g-KFR approaches perform exactly the same time integrals involved in the rate constant calculation while a pole approximation is used in the Keldysh theory. This difference results in the rate constants from the KFRs one to three orders of magnitude lower than that from the Keldysh. Thus, the KFRs give more reliable results for higher intensity range. Furthermore, for the second ionization rate constants of cyclopentanone using the hydrogen-like model with $2p_z$ orbital show the similar trends, i.e., the results from the Keldysh theory may be overestimated and are larger than those from the ADK and the KFR. In Figs. 3(c) and 3(d), the g-KFR results for this double ionization of the molecules are given by the dashed lines.

The Keldysh parameters γ calculated illustrates that, in the region of laser intensity $10^{13} \sim 10^{14} \, W/cm^2$, MPI process dominates both for $\lambda = 394 \, nm$ and $788 \, nm$, even that tunneling ionization plays a role as the ADK contribution to molecular ionization which is not negligible in this laser intensity range. The MPI features can be significantly identified in Fig. 3. First, the values of theoretical slopes for both wavelengths are very close to the minimum number of photons required for MPI of cyclopentanone molecules (the ionization potential is $9.28 \, eV$) experimentally, $2.9 \sim 3.4$ and $5.5 \sim 6.3$, respectively in the $394 \, nm$ and $788 \, nm$ cases, except for the high intensity region where the obvious difference of the calculated curve from the experimental measurement is observed at $788 \, nm$ (Fig. 3d), implying the ionization saturation and heavy fragmentation. Second, the first ionization rate constants of cyclopentanone for $394 \, nm$ are higher than those for 788 nm under the laser intensity below 10^{14} W/cm^2 since the ionization probability of cyclopentanone under the 394 nm irradiation (requiring less photons) is larger than that under the 788 nm irradiation (requiring more photons). Finally, as laser intensity continues to increase up to 10^{15} W/cm^2, the ionization rate constants become independent of the wavelength, as a result that the field ionization becomes the dominant process since, for tunneling ionization or field ionization in intense laser fields, only the initial and final states of the electron are significant while the intermediate states place no rule. However, in some cases an obvious wavelength dependence of molecular ionization process has been observed, that is assigned as collective effect in excitation process of polyatomic molecules.

In pyrrolidine (C_4H_9N) (Ref. 44), ionization rate constants calculated for 800 nm fs-laser fields of $10^{13} \sim 10^{14}$ W/cm^2 by the g-KFR shows a dependence of I^5 roughly in log–log scale for the parent ion yield on the laser intensity below the saturation intensity, a discrepancies in an I^4 power law of the experimental observation. This measured slope of 4 is significantly lower than the minimum number of the photons required for ionizing the molecules, i.e., 6 for pyrrolidine as its ionization potential is 8.77 eV. It implies that the ionization might occur partially through some excited states of neutral pyrrolidine which are characterized by $\sim I^4$ dependent collective excitations as a rate-determining step on the whole ionization process. For this interest an investigation is carried out by quantum chemical calculation for the features of excited states and the photoabsorption of neutral pyrrolidine and the results show a considerably strong absorption band near 200 nm, corresponding to 4-photon absorption in 800 nm laser fields. As mentioned before, since the ionizations through excited-states are totally neglected in the theory like g-KFR, a modified calculation by Floquet method was used for dealing with this excited-state involved ionization. The calculation gave a slope of 4~5 before the saturation, a reasonable agreement with the experimental observations (see Fig. 4). The result indicates that collective excitation process may be worth further considering both in experimental and theoretical methods.

The methods discussed above mainly depend on the accuracy of the ion yields as a function of laser intensity I. As mentioned in the beginning of this section, to measure the laser intensity I accurately is very difficult

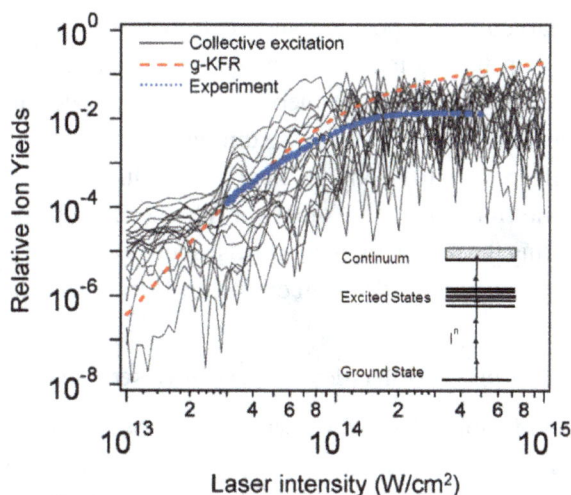

Fig. 4. (Color online) Comparison of the parent ion yields predicted result by using collective excitation model (gray lines, including 30 excited states of the pyrrolidine molecule) and g-KFR theory (red broken line) with that measured by experiment (blue dots). The laser wavelength is 800 nm. The slopes of the results from 30 excited states can be roughly estimated as $4\sim5$ (Ref. 44).

since the size of effective interaction zone changes when the laser output varies in a large intensity region and there are uncertainties from the measurement of the laser parameters such as power, geometry, pulse length and the spatial and temporal profile. These factors often bring a 50% uncertainty in the intensity and thus *in-situ* calibration for intensity I is very important for a precision comparison with the theory. Several methods have been developed with the help of simple strong-field theoretical estimation. From ADK theory the prediction of the saturation intensity, especially for rare gas atoms, is fitted very well with the experimental measurement of tunneling ionization of Xe atom,[15] which has been suggested to be applied for scaling the laser intensity.[43,45,46] Other *in-situ* calibration methods developed mainly depend on the measurement for photoelectron and photoion momentum distribution.[47,48,49] The well-verified *in-situ* intensity calibration is performed by assigning the theoretical predicated "plateau" cutoff energy of $10U_p$ to that of the observed high-order ATI electron spectrum, for example, the high-order ATI spectrum of Xe coming from the electron rescattering process induced by the linearly infrared laser pulse shows very marked cutoff position and clear intensity dependence in the region lower than the saturation intensity, which make the assignments accurate and consistent with the theoretical energy spectrum. Similarly, *in-situ* calibration is feasible for intense ultraviolet laser pulse by assigning the energy shift (related to) from the theoretical value of the low-order ATI spectrum. In additional, there are other ways to measure the peak intensity, such as measuring the proton momentum distributions from H_2.[49] The value of the peak intensity can be determined within 10% in its accuracy.[47]

1.3. Fragmentation of Molecules in Intense Laser Fields

Fragmentation of molecules has shown a rich complexity in intense laser fields. In intense $10 \sim 10^2$ fs-laser pulses, molecular decomposition occurs through various channels that can be from the neutral or charged species and within or after laser irradiation. Dissociation of a molecule can also happen at the nearly same time as its ionization irradiated by an intense laser pulse, i.e., dissociative ionization, a typical case of non Born–Oppenheimer approximation process in which the excitation energy is shared both in electronic and vibrational degrees of freedom.[50] It is very valuable to understand these processes for such applications as controlling fragment ratio and "soft" ionization of large molecules.[51]

1.3.1. *Ionization-dissociation of molecules in intense laser fields and statistical theoretical description*

In many cases of polyatomic molecules, it is assumed that ionization followed by dissociation is the main reaction channel since the rate of the electron motion is faster than that for the nuclear motion. The neutral dissociation requires a relative long time to take place (predissociation or decomposition after internal conversion) after absorbing the energy from laser fields. This timescale is normally longer than 10 fs. Therefore, ionization is the primary process for molecules in intense laser fields. After ionization, dissociation of molecular ions can be subdivided into two categories accordingly whether it occurs within or outside the laser pulse duration. Under intense laser fields, the potential surfaces of molecular ions can be modified by the laser electric field and further induce the ion dissociation during the laser duration. This type of ion dissociation is usually referred to as field assisted dissociation (FAD).[52] The second type of ion dissociation takes place outside the temporal influence of fs-laser pulse. This often happens in the case of polyatomic molecules in which the ions become hot and unstable, by accumulating certain energy from photoexcitation into the vibrational degree of freedom, and finally decomposite. Thus, in theoretical treatment for the molecular dissociation these two categories of dissociation differ on the account of whether such dissociation can or cannot be described in terms of statistical theory.[53]

1.3.2. *Effects of cation absorption on molecular dissociation*

In a "moderate" laser intensity range, molecular ionization can be caused through different mechanisms. Though multiphoton ionization cannot be excluded, field ionization of polyatomic molecules can play a role with an increasing rate constant in this relative "weak" field, for example, the field ionization is dominant, rather than the multiphoton process, for benzene in a laser field of 3.8×10^{13} W/cm^2.[9]

After this first ionization, the produced parent ions is able to gain more excitation energy by further photoabsorption in the same laser pulse for sequential steps, either decompositing or secondary ionizing.[53,54] Thus, it is expected that a role of cation photoabsorption may be significant in this sequential model of the molecular ionization/dissociation in intense laser fields with several 10 fs pulse duration.

The effect of molecular cation resonance was clearly observed in several polyatomic molecules such as aromatic, hydrocarbons, and cycloketones.[37,55−58] Figure 5 gives the mass spectra from a series of cycloketone molecules produced in a 90 fs laser field with 6×10^{13} W/cm^2 (Ref. 37). Table 1 shows the data for the different photoabsorptivities of these molecular cations and the ratios of the parent ion yield to the total ion yield (P^+/T^+) at both the laser wavelengths used, correspondingly. The quite different characters exhibited in the mass spectra for different cycloketones can be explained by the different photoabsorptivities of these molecular cations. It appears that there is a qualitative agreement between absorbency and degree of fragmentation as shown in the case of ketones. It is obvious that the appearance of a parent ion peak is related to the

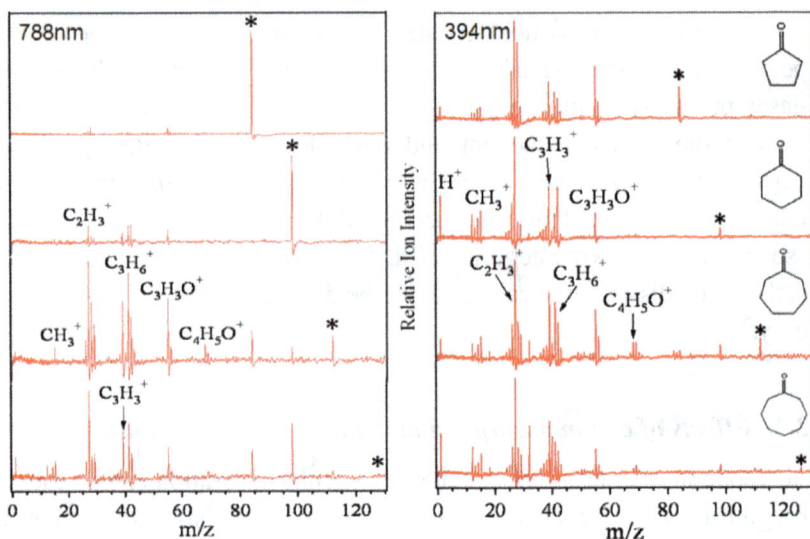

Fig. 5. Mass spectra of cycloketone molecules at 788 nm (left) or 394 nm (right) laser field with the intensity of 6×10^{13} W/cm^2. Form the top to bottom: cyclopentanone, cyclohexanone, cycloheptanone, and cyclooctanone. The peaks of parent ions are denoted by asterisks.[37]

Table 1. Photoabsorption of cyclokenones cations calculated and their corresponding ratios of parent ion yield to total ion yield measured at both 788 and 394 nm laser wavelengths (the calculation is done by using B3LYP method with 6-31++G(d, p)). From Ref. 37.

	λ (nm)	Molar absorptivity (mol^{-1}Lcm^{-1})	P^+/T^+ (%)
Cyclopentanone	394	520.17	5.33
	788	<0.01	81.6
Cyclohexanone	394	570.56	1.62
	788	6.74	52.6
Cycloheptanone	394	2846.74	2.5
	788	880.10	5.5
Cyclooctanone	394	429.20	0.75
	788	3139.13	0.1

absence of the resonant absorption of a single photon by the cations and the fragmentation of the molecular cations increases significantly in the case of resonance.

The resonance effects in these mass spectra were interpreted by the absorption spectra calculated from taking the optimized ionic molecular structures.[37] The optimization is resulted from a nuclear rearrangement from the equilibrium structure, mainly contributed by the H atom movement. Yamanouchi *et al.* have demonstrated that this kind of H migration in organic molecules is as fast as several tens of femtoseconds.[59,60] And therefore within a pulse duration of fs-laser the performation of this rearrangement can be achieved.

1.4. Dissociative Ionization and Coulombic Explosion of Molecules in Intense Laser Fields

When a neutral molecule is exposed to an intense, ultrashort laser pulse, the molecule is first ionized and then a subsequent fragmentation occurs.[61,62] For single ionization, low kinetic energy ion fragments are observed but for multiply ionization Coulomb explosion produces energetic fragment ions.[63,64] For a Coulomb explosion reaction of molecular AB into fragment ions $A^{p+} + B^{q+}$, the total kinetic energy in the (p, q) channel, $E(p, q)$, can be calculated simply by

$$E(p, q) = \frac{pqe^2}{4\pi\varepsilon_0 R} = 14.4\frac{pq}{R} \text{ (eV)}, \qquad (1.27)$$

where the repulsion is assumed to be due to point charges located in each of the fragments. With Eq. (1.27) and the calculated ratio for energies shared in two different mass fragments involved, the molecular equilibrium distance R can be determined from the measured kinetic energy released (KER) of each fragment.

The ion fragment following the dissociative ionization contains abundant information. The angular distributions of those ions can be affected by many factors, such as the laser intensity, pulse width and the symmetry of the highest occupied molecular orbital (HOMO), multi-nuclear interference and so on[38,65] In particular, angle-dependent ionization rates which reflect the symmetry of the HOMO, and dynamic alignment of the neutral molecule by intense laser pulses must be considered[66,67] to explain the anisotropy of the ionization process, which affects the angular distributions of the fragment ions. Thus, with the help of ultrashort laser and momentum or angle-resolved kinetic energy measurement, in principal this method enables us to investigate molecular structure and multi-particle correlated dynamics in intense laser fields.[68] However, our knowledge is quite limited till now for larger polyatomic molecules since these molecules are a multielectron system and have large electronic orbitals over the multicenter of molecules.

1.4.1. *Dissociative ionization of formic acid molecules*

The hydrogen emissions and hydrogen migrations play important roles in the molecule decomposition and rearrangement dynamics because they move on a time-scale faster than most of other ion fragmentation processes.[69] The hydrogen migration channels of many organic molecules have been observed and the time scale of the hydrogen migrations have been discussed.[70–72]

As the simplest carboxylic acid, formic acid (HCOOH) attracted many researches. The ionic dissociation channels of formic acid have been studied by vacuum-ultraviolet (VUV) spectroscopy. The ground state $1^2A'$ (IE = 11.33 eV) of singly charge ion is stable and its first excited ion state $1^2A''$ (IE = 12.37 eV) is associated with two dissociation channels, $COOH^+ + H$ and $HCO^+ + OH$. For other lower excited state, $2^2A'$ (IE = 14.81 eV), $2^2A''$ (IE = 15.75 eV) and $3^2A'$ (IE = 16.97 eV), the $HCO^+ + OH$ channel is also important. However, the $HCO + OH^+$ channel is only open for excited states higher than the $3^2A'$ state. In intense laser fields, the formic acid can decompose to several ion fragments [Wang *et al.*, 2010]. Parent ions and fragments obtained by the loss of one H atom are dominant. The main fragments are $COOH^+$, COO^+, $HCOH^+$, HCO^+, H_2O^+, OH^+ and O^+ from various dissociation channels. To identify these channels, information such as mass, kinetic energy released, angular distribution of these fragments produced, and even the correlation between the fragments are necessary. Therefore, a velocity map imaging apparatus with a 100 *fs* pulsed 800 nm laser of $\sim 10^{15}$ W/cm^2 is employed for this study.

For singly charged parent ions, the experimental measurement suggests that the dissociation occurs with the channel of HCOOH → $HCO^+ + OH$

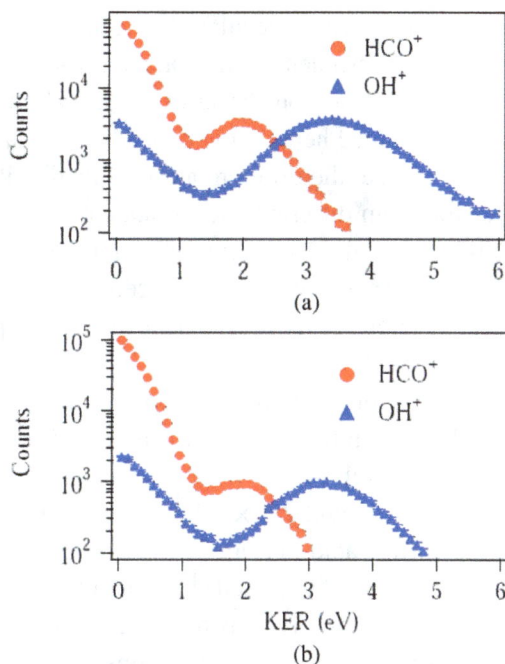

Fig. 6.   KERs of HCO^+ and OH^+ at 2.4×10^{14} W/cm^2 (a) and 9×10^{13} W/cm^2 (b) with a linear polarized laser at 800 nm.

$+ e^-$, from the excited states $1^2A''$, $2^2A'$, $2^2A''$ and $3^2A'$ of the parent ion $HCOOH^+$, and the channel of $HCOOH \rightarrow HCO + OH^+ + e^-$, mainly from the $3^2A'$ state. Both channels lead to the dissociation of the same C–O single bond. The KERs of these fragment ions show clearly bimodal kinetic energy distributions (Fig. 1.6), suggesting that besides the dissociation channels described above the high-energy channels exist, i.e., from Coulomb explosion of double charged parent ions as discussed later in this section. Additional, the counts of low energy OH^+ are much less than those of low energy HCO^+, consistent with that OH^+ can be produced by the dissociation only from the higher excited state $3^2A'$ while HCO^+ may come from lower excited states.

Angular distributions of these fragments give the maxima along the laser polarization. These anisotropic angular distributions of fragmental ions commonly can be attributed to dynamic alignment, the molecular axis aligned along the laser polarization due to the interaction of laser electric field and the induced molecular dipole moment, or geometric alignment with the angle dependent ionization rate. Normally, the dynamic alignment shows a dependence on the laser intensity.[67] Because all the angular distributions obtained here do not depend on the laser intensity in the present intensity regime, the different anisotropic distributions HCO^+ and OH^+, dissociated from different excited state of the parent ion, may origin from the different angle-dependent ionization rate.

The double charged parent ions are produced with a significant yield in the 800 nm fs-laser intensity used. It is believed that sequential double ionization plays a dominate role in this case, as the data show that the circularly polarized laser pulse does not suppress the ion pair yield and the KER does not depend on the laser polarization and intensity, too. The coincidence maps of the TOF of fragment ions are shown in Fig. 7 at the laser intensities of 9×10^{13} and 2.4×10^{14} W/cm². These maps indicate that doubly charged formic acid ions are produced and decomposed from the COH^+–OH^+, COH^+–O^+, CO^+–OH^+, and CO^+–O^+ ion pairs. The H_2O^+–CO^+ ion pair is also observed as result of fast hydrogen migration with a low yield. These experimental observations of ion–ion coincidence momentum indicate that the two-body dissociative ionization process of doubly charged formic acid ($HCOOH^{2+}$) is induced by intense fs-laser fields in the present intensity regime. Thus, the main channels of dissociative

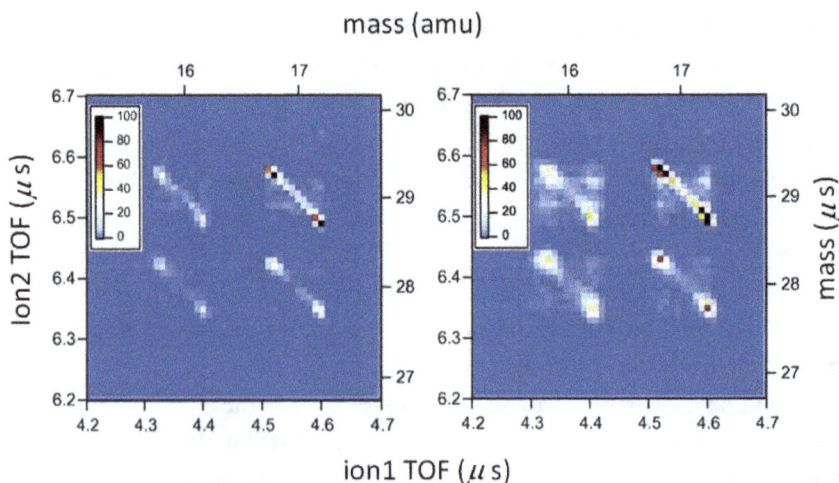

Fig. 7. Ion–ion coincidence map at the intensities of 9×10^{13} W/cm² (left) and 2.4×10^{14} W/cm² (right) with linearly polarized 800 nm laser.[39]

ionization identified are follows: (1) HCOOH $\rightarrow$ OH$^+$+ COH$^+$+ 2e^-, (2) HCOOH $\rightarrow$ O$^+$+COH$^+$ + H + 2e^-, (3) HCOOH $\rightarrow$ OH$^+$+ CO$^+$+ H + 2e^-, and (4) HCOOH $\rightarrow$ O$^+$+ CO$^+$+ 2H + 2e^-.

The Coulomb explosion channels with one neutral hydrogen emission have been identified using the ion–ion coincidence method. The different effective ionization potential of fragment ions and structure of precursor states have been discussed and are believed to arouse the different KER distributions. Another two weak two-body Coulomb explosion channels have been identified. One is from the hydrogen migration and the other is from the two-body Coulomb explosion of C = O double bond.

1.4.2. *Coulombic explosion of CH$_3$I*

As a typical result of multiple charged parent molecules produced by the interaction with intense laser fields, the parent molecules will fragment into small positively charged species, in most of cases, due to a Coulomb explosion of the parent molecules. The kinetic energy disposed into a fragment ion can be estimated by the stationary position just before the onset of the Coulomb explosion. But, usually this energy is found to be substantially lower than the one expected for a prompt dissociation of the charged parent molecules at the equilibrium bond length R_e. It was inferred that a Coulomb explosion typically occurs at a larger critical bond length R_c than R_e due to the bond-softening induced by the laser field.[73] The temporal rise of the intensity of laser pulse brings about a weakening of the binding energy of the valence electrons to the molecules, consequently the molecular bond length q relaxes from its equilibrium length R_e until the laser pulse reaches its maximum intensity at which the bond length stabilizes as its critical length R_c. Hence the molecule ionizes predominantly at $R \approx R_c$. This may give a good measure of molecular bond length and be interesting for developing a way of quantum tomograph for molecular structures.

Liu *et al.*[74] studied the ionization–dissociation dynamics of CH$_3$I with a time-of-flight mass spectrometer and assigned the fragments CH$_3^{p+}$ and I^{q+} to different multiphoton dissociative ionization and Coulomb explosion channels. Coulomb explosion of CH$_3$I was found to occur at a value of $\sim$3.7 Å. Subsequently, Wang *et al.*[75] applied the velocity map imaging method to explore the recoil of the ionic fragments produced from CH$_3$I under 35 fs 800 nm laser field. They found that at low laser intensity the recoil velocity distributions of CH_3^+ and I^+ are solely produced from multiphoton dissociative ionization, while at the more intense laser field (6×10^{14} W/cm^2) Coulomb explosion channels are included to form I^+, I^{2+} and I^{3+} fragments. Most recently, Corrales *et al.*[76] studied the Coulomb explosion of CH$_3$I in an intense 50 fs 804 nm laser field with velocity map imaging and one-dimensional wave packet calculation. Their results reveal the existence of a potential energy barrier due to a bound which is minimum in the potential energy curve of the CH$_3$I^{2+} species and a strong stabilization with respect to Coulombic repulsion for the higher charged CH$_3$I^{n+} ($n = 3, 4$) species.

We investigated the reaction of $CH_3I^{n+} \rightarrow CH_3^{p+} + I^{q+}$ ($n = p + q$) in a 100 fs 800 nm linearly polarized laser fields of $2.6 \times 10^{14} \sim 5.8 \times 10^{14}$ W/cm^2 by means of a velocity map imaging method. A typical kinetic energy distribution of the various atomic fragment ions I^{q+} ($q = 1-3$) is given in the column of Fig. 8 for a laser intensity of 3.8×10^{14} W/cm^2. Here

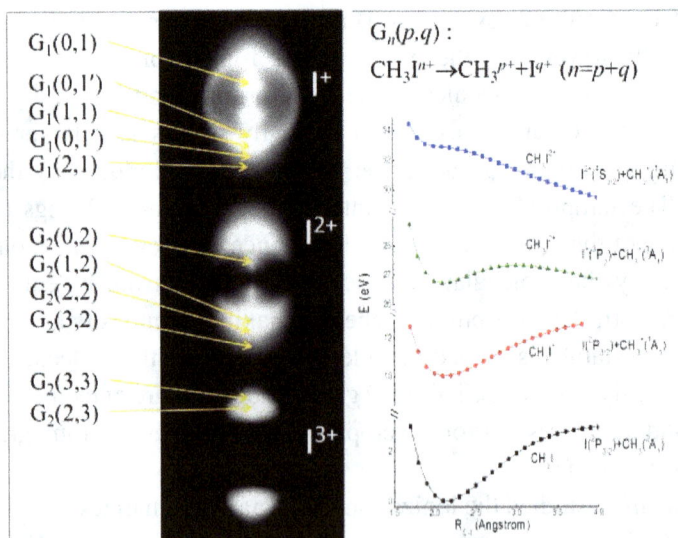

Fig. 8.   Velocity map imaging of I^+, I^{2+} and I^{3+} from Coulomb explosion of CH$_3$I in the intense laser field of 3.7×10^{14} W/cm^2. Some of the reaction channels are labeled in the images and the KERs are obtained from the two-dimension distribution of the product ionic fragments. The left is the calculated potential curves for neutral and charged CH$_3$I.

the polarization vector of the laser is parallel to the vertical direction of the image plane. These images show that each I^{q+} exhibits a strong preference to recoil along the polarization vector of the intense *fs*-laser field. The kinetic energy distribution of the atomic fragment ions I^{q+} can be obtained from these velocity map images by a fit of multiple Gaussian functions and several dissociative ionization and Coulomb explosion channels have been identified for I^{q+}.[38]

The measured total kinetic energy of a Coulomb explosion channel, $G_n(p, q)$, is much lower than Coulomb repulsion energy calculated by Eq. (1.28) with an equilibrium C–I internuclear distance R_e of 2.14 Å. For example, the calculated kinetic energy of channel $G_2(1, 2)$ is 13.46 eV, while the measured total kinetic energy is only 7.7 eV. This could be well explained by enhanced ionization at the critical internuclear distance, R_c, (much larger than R_e), where the sequential ionization and Coulomb explosion occur. Enhanced ionization has a strong angular dependence, resulting in observed strongly anisotropic angular distribution of the products from the Coulomb explosion channels. The critical internuclear distance R_c can be calculated from the experimentally measured kinetic energy of the I^{q+}. From the measured kinetic energy from Coulomb explosion channels $G_2(1, 2)$ and $G_1(2, 1)$, the reduced internuclear distance is about 3.8 Å for CH_3I^{3+}, this result is consisted with the value of 3.7 Å reported by Sugita *et al.*[77] and Wang *et al.*[75]

1.5. Summary and Perspectives

In this chapter, we have reported experimental and theoretical results on the dynamics of polyatomic molecules induced by intense fs-laser fields of $10^{12} \sim 10^{14}$ W/cm^2. Emphases have been paid on the ionization rates, resonance effects of molecular cation and dissociative ionization or Coulomb explosion with some molecules as example. The studies show some of basic features and dynamic processes of polyatomic molecules induced by intense laser pulses, summarized as follows:

(1) Multiphoton ionization or tunneling ionization is the first and the most important step in molecular processes induced by intense fs-laser pulses.

(2) Understanding of molecular ionization rates plays an important role in exploring the pictures of mechanics and developing the theoretical description for interaction of molecule-intense laser fields.

(3) The behavior of molecules induced by intense fs-laser fields can be affected significantly by photoaborption of molecular cations within the fs-laser pulse.

(4) Dissociative ionization or Coulomb explosion induced by intense fs-laser fields provides a new chance for exploring electronic-vibrational coupling and obtaining molecular tomograph.

As a fast growing field, the studies on molecular dynamics in intense fs-laser fields bring us many fantastical findings continuously. The major breakthrough can be expected in many subjects and we only list several of them here:

- *Coherent control of molecular processes induced by* fs-*laser pulses*
  — to tailor fs-laser pulse shape and polarization for inducing coherent coupling of reaction pathways and optimizing molecular processes

- *Ultrafast dynamics of molecular dissociation induced by* fs-*laser fields*
  — to attract the attention to the nonperturbative nature of the laser field coupled with the interplay of the electronic and nuclear degrees of freedom and the *rescattering* feature from interaction of electron- or photon-molecules.

- *CEP stabilized few-cycle* fs-*laser pulse and its application in quantum control of molecular ionization and dissociation*
  — to study laser-molecule interaction in a qualitatively new region where CEP becomes an important parameter and thereby controls the dynamics of the much faster electrons within sub-fs pulses

Many puzzles remain in the dynamics of molecules in intense fs-laser fields. Though the plausibility of some methods, both theoretical and experimental, has been demonstrated, more detailed studies need to be performed, and an understanding of the dynamics in the electron move time scale and atom dimension has yet to be found. All these will become one of the most productive fields of atomic, molecular and optical physics for sure in the near future.

Acknowledgments

The work was supported by National Basic Research Program of China (973 Program) under grant No. 2013CB922200 and by National Science Foundation of China under grants No. 11034003, 11127403, and 10534010. We would like to thank Shen-Hseng Lin, Kiyoshi Ueda, for illuminating and helpful discussions, and many collaborators involved in the studies during years.

References

1. G. Mourou, T. Tajima and S. V. Bulanov, *Rev. Mod. Phys.* **78**, 309 (2006).
2. E. Goulielmakis, E. M. Schultze, M. Hofstetter, V. S. Yakovlev, J. Gagnon, M. Uiberacker, A. L. Aquila, E. M. Gullikson, D. T. Attwood, R. Kienberger, F. Krausz and U. Kleineberg, *Science* **320**, 1614 (2008).
3. J. H. Posthumus, *Rep. Prog. Phys.* **67**, 623 (2004).
4. S. L. Chin and P. Lambropoulos, *Multiphoton Ionization of Atoms* (Academic Press, Orlando, 1984).
5. L. V. Keldysh, *Zh. Eksp. Teor. Fiz.* **47**, 1515 (1964); *Sov. Phys. JETP* **20**, 1307 (1965).
6. D. M. Volkov, *Z. Phys.* **94**, 250 (1935).
7. S. L. Chin, F. Yergeau and P. Lavigne, *J. Phys. B* **18**, L213 (1985).
8. P. B. Corkum, *Phys. Rev. Lett.* **71**, 1994 (1993).
9a. M. J. DeWitt and R. J. Levis, *Phys. Rev. Lett.* **81**, 5101 (1998).
9b. M. J. DeWitt and R. J. Levis, *J. Chem. Phys.* **108**, 7045 (1998).
10. K. J. Schafer, B. Yang, L. I. DiMauro and K. C. Kulander, *Phys. Rev. Lett.* **70**, 1599 (1993).
11. M. Lewenstein, Ph. Balcou, M. Y. Ivanov, A. L'Huillier and P. B. Corkum, *Phys. Rev. A* **49**, 2117 (1994).
12. A. Stolow, A. E. Bragg and D. M. Neumark, *Chem. Rev.* **104**, 1719 (2004).
13. M. Wollenhaupt, V. Engel and T. Baumert, *Annu. Rev. Phys. Chem.* **56**, 25 (2005).
14. F. Krausz and M. Ivanov, *Rev. Mod. Phys.* **81**, 163 (2009).
15. T. D. G. Walsh, F. A. Ilkov, J. E. Decker and S. L. Chin, *J. Phys. B: At. Mol. Phys.* **27**, 3767 (1994).
16. X. M. Tong, Z. X. Zhao and C. D. Lin, *Phys. Rev. A* **66**, 033402 (2002).
17. E. Wells, M. J. DeWitt and R. R. Jone, *Phys. Rev. A* **66**, 013409 (2002).
18. K. J. Schafer and K. C. Kulander, *Phys. Rev. A* **42**, 5794 (1990).
19. H. G. Muller and F. C. Kooiman, *Phys. Rev. Lett.* **81**, 1207 (1998).
20. H. R. Reiss, *Prog. Quantum Electron.* **16**, 1 (1992).
21. H. R. Reiss, *Phys. Rev. A* **42**, 1476 (1990).
22. A. M. Perelomov, V. S. Popov and M. V. Terentev, *Sov. Phys. JETP.* **23**, 924 (1966).
23. M. V. Amomosov, N. B. Delone, and V. P. Krainov, *Sov. Phys. JETP* **64**, 1191 (1986).
24. V. I. Usachenko and S. I. Chu, *Phys. Rev.* **71**, 063410 (2005).
25. K. Mishima, M. Hayashi and S. H. Lin, *Phys. Rev. A* **71**, 053411 (2005).
26. J. Muth-Bohm, A. Becker and F. H. M. Faisal, *Phys. Rev. Lett.* **85**, 2280 (2000).

27. H. Mineo, S. D. Chao, K. Nagaya, K. Mishima, M. Hayashi and S. H. Lin, *Chem. Phys. Lett.* **439**, 224 (2007).

28. F. Grossmann, *Theoretical Femtosecond Physics, Atoms and Molecules in Strong Laser Fields* (Springer-Verlag, Berlin, Heidelberg, 2008).

29. F. H. M. Faisal, *J. Phys. B, At. Mol. Phys.* **6**, L89 (1973).

30. H. R. Reiss, *Phys. Rev. A* **22**, 1786 (1980).

31. Q. Wang, D. Wu, M. Jin, F. Liu, F. Hu, X. Cheng, H. Liu, Z. Hu, D. Ding, H. Mineo, Y. A. Dyakov, A. M. Mebel, A. D. Chao and S. H. Lin, *J. Chem. Phys.* **129**, 204302 (2008).

32. M. V. Ammosov, N. B. Delone and V. P. Krainov, *Sov. Phys. JETP.* **64**, 1191 (1986).

33. T. K. Kjeldsen, C. Z. Bisgaard, L. B. Madsen and H. Stapelfeldt, *Phys. Rev. A* **71**, 013418 (2005).

34. C. B. Madsen and L. B. Madsen, *Phys. Rev. A* **76**, 043419 (2007).

35. T. Brabec, M. Cote, P. Boulanger and L. Ramunno, *Phys. Rev. Lett.* **95**, 073001 (2005).

36. Z. X. Zhao and T. Brabec, *J. Mod. Opt.* **54**, 981 (2007).

37. D. Wu, Q. Wang, X. Cheng, M. Jin, X. Li, Z. Hu and D. Ding, *J. Phys. Chem. A* **111**, 9494 (2007).

38. D. Zhang, H. Xu, M. Jin and D. Ding, *J. Phys.: Conf. Ser.* **388**, 032065 (2012).

39. C. Wang, D. Ding, M. Okunishi, Z.-G. Wang, X.-J. Liu, G. Prümper and K. Ueda, *Chem. Phys. Lett.* **496**, 32 (2010).

40. D. H. Parker and A. T. J. B. Eppink, *J. Chem. Phys.* **107**, 2357 (1997).

41. G. Prumper, H. Fukuzawa, T. Lischke and K. Ueda, *Rev. Sci. Instrum.* **78**, 083104 (2007).

42. K. Nagaya, H. Mineo, K. Mishima, A. A. Villaeys, M. Hayashi and S. H. Lin, *Phys. Rev. A* **75**, 013402 (2007).

43. Q. Wang, H. Mineo, D. Wu, M. X. Jin, C. H. Chin, Y. Teranishid, S. D. Chao, D. Ding and S. H. Lin, *Laser Phys.* **19**, 1671 (2009).

44. Q. Wang, D. Wu, D. Zhang, M. Jin, F. Liu, H. Liu, Z. Hu, D. Ding, H. Mineo, Y. A. Dyakov, Y. Teranishi, S. D. Chao, A. M. Mebel and S. H. Lin, *J. Phys. Chem. C* **113**, 11805 (2009).

45. S. M. Hankin, D. M. Villeneuve, P. B. Corkum and D. M. Rayner, *Phys. Rev. A* **64**, 013405 (2001).

46. C. Wang, M. Okunishi, R. R. Lucchese, T. Morishita, O. I. Tolstikhin, L. B. Madsen, K. Shimada, D. Ding and K. Ueda, *J. Phys. B: At. Mol. Opt. Phys.* **45**, 131001 (2012).

47. A. S. Alnaser, X. M. Tong, T. Osipov, S. Voss, C. M. Maharjan, B. Shan, Z. Chang and C. L. Cocke, *Phys. Rev. A* **70**, 023413 (2004).

48. C. Smeenk, J. Z. Salvail, L. Arissian, P. B. Corkum, C. T. Hebeisen and A. Staudte, *Opt. Exp.* **19**, 9336 (2011).

49. D. Ray, Z. Chen, S. De, W. Cao, I. V. Litvinyuk, A. T. Le, C. D. Lin, M. F. Kling and C. L. Cocke, *Phys. Rev. A* **83**, 013410 (2011).

50. J. A. Davies, R. E. Continetti, D. W. Chandler and C. C. Hayden, *Phys. Rev. Lett.* **84**, 5983 (2000).

51. S. M. Hankin, D. M. Villeneuve, P. B. Corkum and D. M. Rayner, *Phys. Rev. Lett.* **84**, 5082 (2000).
52. X. P. Tang, S. F. Wang, M. E. Elshakre, L. R. Gao, Y. L. Wang, H. F. Wang and F. A. Kong, *J. Phys. Chem. A* **107**, 13 (2003).
53. E. W.-G. Diau, J. L. Herek, Z. H. Kim and A. H. Zewail, *Science* **279**, 847 (1998).
54. L. Robson, K. W. D. Ledingham, A. D. Tasker, P. McKenna, T. McCanny, C. Kosmidis, D. A. Jaroszynski, D. R. Jones, R. C. Issac and S. Jamieson, *Chem. Phys. Lett.* **360**, 382 (2002).
55. R. Itakura, J. Watanabe, A. Hishikawa and K. Yamanouchi, *J. Chem. Phys.* **114**, 5598 (2001).
56. H. Harada, S. Shimizu, T. Yatsuhashi, S. Sakabe, Y. Izawa and N. Nakashima, *Chem. Phys. Lett.* **342**, 563 (2001).
57. S. A. Trushin, W. Fuß and W. E. Schmid, *J. Phys. B: At. Mol. Phys.* **37**, 3987 (2004).
58. T. Yatsuhashi and N. Nakashima, *J. Phys. Chem. A* **114**, 7445 (2010).
59. R. Itakura, P. Liu, Y. Furukawa, T. Okino, K. Yamanouchi and H. Nakano, *J. Chem. Phys.* **127**, 104306 (2007).
60. T. Okino, A. Watanabe, H. Xu and K. Yamanouchi, *Phys. Chem. Chem. Phys.* **14**, 10640 (2012).
61. V. R. Bhardwaj, D. M. Rayner, D. M. Villeneuve and P. B. Corkum, *Phys. Rev. Lett.* **87**, 253003 (2001).
62. M. Castillejo, S. Couris, E. Koudoumas and M. Martín, *Chem. Phys. Lett.* **308**, 373 (1999).
63. L. J. Frasinski, K. Kodling and P. Hatherly, *Phys. Rev. Lett.* **58**, 2424 (1987).
64. A. Hishikawa and K. Yamanouchi, Coulomb explosion imaging of molecular dynamics in intense laser fields, Chapter 1, In *Progress in Ultrafast Intense Laser Science* II, eds. K. Yamanouchi, S. L. Chin, Agostini, P. and G. Ferrante (Springer-Verlag, Berlin Heidelberg, 2007), pp. 1–24.
65. X. Zhang, D. Zhang, H. Liu, H. Xu, M. Jin and D. Ding, *J. Phys. B: At. Mol. Opt. Phys.* **43**, 025102 (2010).
66. I. V. Litvinyuk, F. L. Kevin, P. W. Dooley, D. M. Rayner, D. M. Villeneuve and P. B. Corkum, *Phys. Rev. Lett.* **90**, 233003 (2003).
67. H. Stapelfeldt and T. Seideman, *Rev. Mod. Phys.* **75**, 543 (2003).
68. V. L. B. de Jesus, A. Rudenko, B. Feuerstein, K. Zrost, R. Schröter, C. D. Moshammer and J. Ullrich, *J. Electron Spectr. Related Phenom.* **141**, 127 (2004).
69. S. Roither, X. Xie, D. Kartashov, L. Zhang, M. Schöffler, H. Xu, A. Iwasaki, T. Okino, K. Yamanouchi, A. Baltuska and M. Kitzler, *Phys. Rev. Lett.* **106**, 163001 (2011).
70. H. L. Xu, T. Okino and K. Yamanouchi, *Chem. Phys. Lett.* **469**, 255 (2009).
71. T. Okino, Y. Furukawa, P. Liu, T. Ichikawa, R. Itakura, K. Hoshina, K. Yamanouchi and H. Nakano, *Chem. Phys. Lett.* **419**, 223 (2006).
72. A. Hishikawa, A. Matsuda, M. Fushitani and E. J. Takahashi, *Phys. Rev. Lett.* **99**, 258302 (2007).
73. S. Chelkowski, A. D. Bandrauk, A. Staudte and P. B. Corkum, *Phys. Rev. A* **76**, 013405 (2007).
74. H. Liu, Z. Yang, Z. Gao and Z. Tang, *J. Chem. Phys.* **126**, 044316 (2007).

75. Y. Wang, S. Zhang, Z. Wei and B. Zhang, *J. Phys. Chem. A* **112**, 3846 (2008).
76. M. E. Corrales, G. Gitzinger, J. Gonzáalez-Váazquez, V. Loriot, R. de Nalda and L. Bañares, *J. Phys. Chem. A* **116**, 2669 (2012).
77. A. Sugita, M. Mashino, M. Kawasaki, Y. Matsumi, R. J. Gordon and R. Bersohn, *J. Chem. Phys.* **112**, 2164 (2000).

CHAPTER 4

ULTRAFAST INTERNAL CONVERSION OF PYRAZINE VIA CONICAL INTERSECTION

T. Suzuki[*,†] and Y. I. Suzuki[*]

We describe recent experimental studies of internal conversion via conical intersection in pyrazine. Ultrafast $S_2 - S_1$ internal conversion is observed in real time using a time-resolved photoelectron imaging (TRPEI) method with a time resolution of 22 fs. This method enables us to obtain a time-energy map of the photoelectron angular anisotropy, which unambiguously reveals the signature of internal conversion. Furthermore, the time-energy map of the photoelectron kinetic energy distribution (PKED) exhibits vibrational quantum beats of totally symmetric modes in S_1 after internal conversion. We also studied similar conical intersections between $D_1(\pi^{-1})$ and $D_0(n^{-1})$ by He(I) photoelectron spectroscopy and pulsed field ionization photoelectron spectroscopy. The existence of ultrafast internal conversion from D_1 to D_0 is confirmed by broadening of He(I) photoelectron spectra of pyrazine and deuterated pyrazine. Comparison of these spectra with one-color resonance-enhanced multiphoton ionization (REMPI) spectra of the 3s Rydberg states clearly indicates that the conical intersection between D_1 and D_0 induces ultrafast internal conversion from the Rydberg state with a D_1 ion core to that with a D_0 ion core.

1.1. Introduction

Employing the Born–Oppenheimer approximation, photophysical and photochemical processes are treated as classical trajectories or

---

*Department of Chemistry, Graduate School of Science, Kyoto University, Kyoto 606-8502, Japan
†RIKEN Center for Advanced Photonics, RIKEN, Wako, Saitama 351-0198, Japan

quantum-mechanical wave packet motions of nuclear geometry on potential energy surfaces.[1] However, the approximation breaks down when potential energy surfaces become energetically close to each other, which permits nonadiabatic transitions to occur between potential energy surfaces. In a diatomic molecule, the potential energy curves of electronic states with the same symmetry give rise to avoided crossings according to the von Neumann–Wigner noncrossing rule.[2] On the other hand, the potential energy surfaces of a polyatomic molecule with N internal degrees of freedom can form a seam of crossings in $N - 2$ dimensional space; such a seam is termed a conical intersection.

Conical intersections are the most important topographic features of multidimensional surfaces that induce nonadiabatic transitions in polyatomic molecules.[4-8] The characteristic conical shape funnels nuclear trajectories on an upper surface to a lower one, facilitating efficient internal conversion (a spin-allowed nonradiative transition). Photodissociation of ammonia[9-12] is a well-known example of a photochemical reaction mediated by a conical intersection. The electronic ground state of ammonia belongs to the C_{3v} point group (the permutation–inversion group is required for a rigorous description), while the first excited singlet state has a trigonal planar D_{3h} minimum. When one of the N–H bonds is elongated in the planar geometry, the symmetry of the system becomes C_{2v} in which intersection of the excited and the ground-state potential energy surfaces is symmetry allowed. However, these two potential energy surfaces avoid each other in the nonplanar geometry because the two electronic states have the same irreducible representation in this distorted geometry (Fig. 1). The minimum energy conical intersection generally plays a crucial role in the dynamics as it determines the accessibility of the nuclear wave packet (or trajectory) to the seam of crossings.

Conical intersections started attracting attention in the late 1960s and the early 1970s as the origins for various photophysical and photochemical processes.[13-15] However, in those days, it was difficult to identify conical intersections in the high-dimensional configuration space of polyatomic molecules. Extensive and accurate computations of potential energy surfaces performed in recent years indicate that conical intersections are ubiquitous in polyatomic molecules.

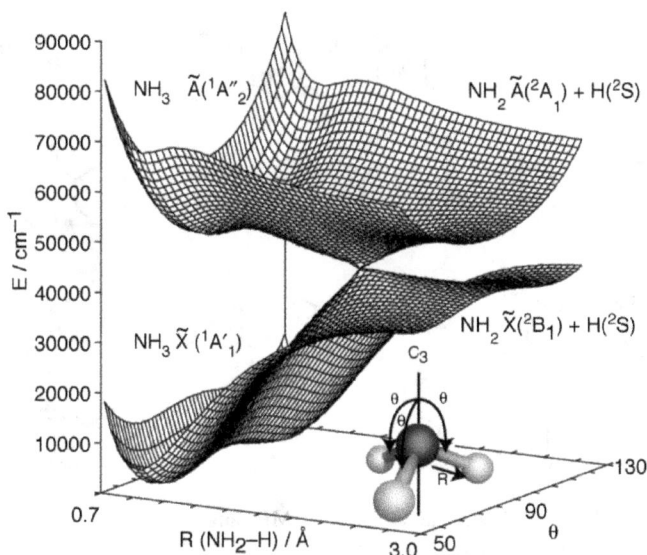

Fig. 1. Potential energy surfaces for the A and X states of ammonia for NH_2–H stretching and out-of-plane bending angle θ. Reproduced with permission from Ref. 3, copyright (2006) by American Institute of Physics.

1.2. Pyrazine: Ultrafast $S_2(^1B_{2u}, \pi\pi^*)$ — $S_1(^1B_{3u}, n\pi^*)$ Internal Conversion Via Conical Intersection

The $S_2(^1B_{2u}, \pi\pi^*)$–$S_1(^1B_{3u}, n\pi^*)$ internal conversion of pyrazine ($C_4H_4N_2$, D_{2h}) is one of the best-known examples of ultrafast electronic deactivation via a conical intersection.[16] The topography of this conical intersection has been extensively studied by *ab initio* molecular orbital calculations.[17-30] Although pyrazine has 24 normal modes (see Fig. 2), only a single mode, $Q_{10a}(b_{1g})$, mediates $S_2 - S_1$ coupling due to the selection rule. Furthermore, only a few totally symmetric (a_g) modes participate in the short-time vibrational dynamics of this system. This reduced dimensionality makes pyrazine a benchmark for theoretical studies of ultrafast internal conversion via a conical intersection. The conical intersection of pyrazine is depicted in Fig. 3 for the two-dimensional space of Q_{10a} and Q_{6a}.[18]

The $S_2 \leftarrow S_0$ photoabsorption spectrum of pyrazine in the deep ultraviolet region, 230–280 nm, exhibits a broad feature, which implies

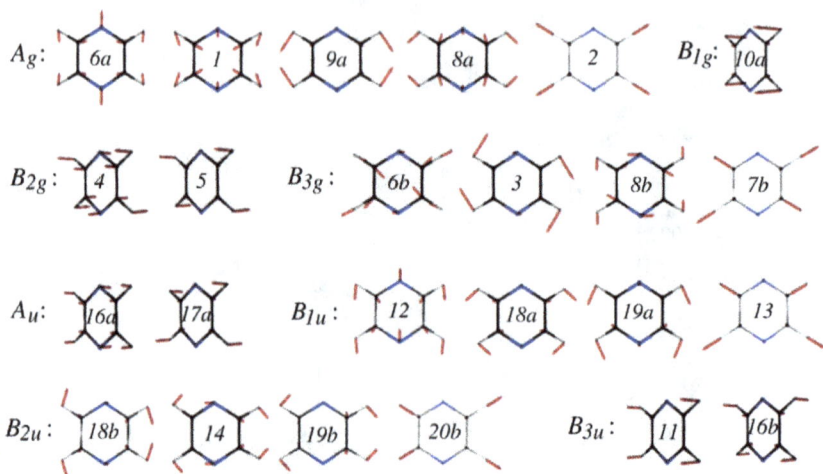

Fig. 2.   Normal modes of pyrazine calculated by MP2/aug-cc-pVDZ level. They are labeled with the Wilson's numbering for benzene.[31,32]

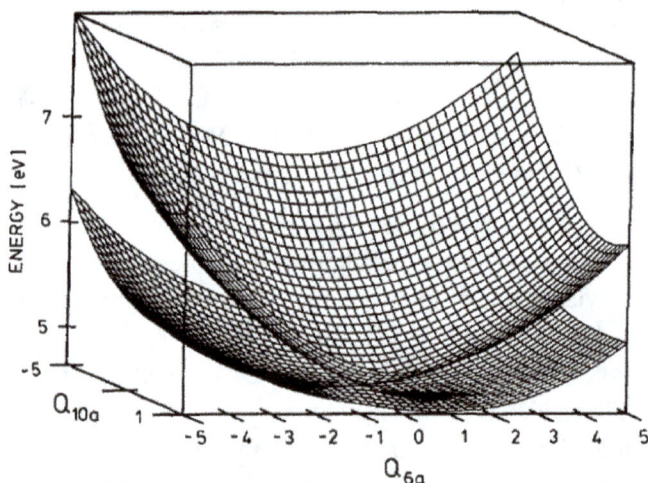

Fig. 3.   Conical intersection of S_2 and S_1 adiabatic potential energy surfaces of pyrazine in the two-dimensional space spanned by Q_{10a} and Q_{6a}. Reproduced with permission from Ref. 18, copyright (1994) by American Institute of Physics.

ultrafast decay of the S_2 state. Figure 4 shows the spectra of pyrazine (pyrazine-h4) and fully deuterated pyrazine (pyrazine-d4) vapor at room temperature.[33] Despite their broadness, some vibrational structures are discernible in these $S_2 \leftarrow S_0$ spectra; for example, the progressions of

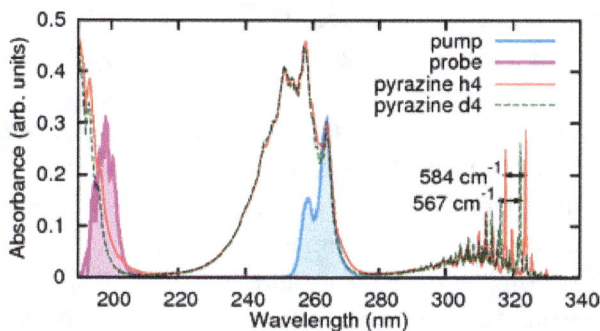

Fig. 4. UV photoabsorption spectra of S_1, S_2, and S_3 of pyrazine-h4 (thin solid line) and pyrazine-d4 (thin dashed line) at room temperature. The spectra of our pump (264 nm, 4.70 eV) and probe (198 nm, 6.26 eV) pulses are also shown by the thick solid lines. Reproduced with permission from Ref. 33, copyright (2010) by American Institute of Physics.

totally symmetric modes, Q_1 and Q_{6a}, are assigned based on comparison of their vibrational frequencies with those in the S_0 state ($\nu_1 = 1014\,cm^{-1}$ and $\nu_{6a} = 596\,cm^{-1}$ for pyrazine-h4).[34] Several theoretical studies have simulated the spectrum of pyrazine-h4; for instance, Woywod *et al.* reproduced the observed spectrum by performing *ab initio* calculations by the complete-active-space self-consistent-field and multireference configuration interaction methods.[18] These studies of the absorption spectrum led to important advances in the understanding of the nonadiabatic dynamics of pyrazine. On the other hand, since an absorption spectrum is the Fourier transform of the autocorrelation function of the wave packet prepared by photoexcitation,[35,36] it provides very limited information about outside the Franck–Condon region.

Seel and Domcke[37] proposed using ultrafast photoelectron spectroscopy to investigate nonadiabatic wave packet dynamics over wide regions containing multiple potential energy surfaces. This technique employs a pump pulse to generate a nuclear wave packet on the S_2 potential energy surface at the time origin ($t = 0$) and it uses a time-delayed probe pulse to interrogate this wave packet by projecting it onto the cationic-state wavefunctions of D_0 ($^2A_g, n^{-1}$) and D_1 ($^2B_{1g}, \pi^{-1}$). To illustrate the principle of this method, Seel and Domcke simulated time-dependent photoelectron kinetic energy distributions (PKEDs) for pyrazine with hypothetical pump [full width at half maximum (FWHM): 1 fs] and probe (16 fs) pulses; the simulated PKEDs revealed a vibrational wave packet oscillating between the S_2 and S_1 diabatic surfaces. Seel and Domcke considered three vibrational modes: Q_1, Q_{6a}, and Q_{10a}. Later, Hahn and Stock[38] included Q_{9a} in their consideration. More recently, Werner *et al.*[29] have performed nonadiabatic molecular dynamics simulations "on the fly" using a time-dependent density functional theory with all vibrational degrees of freedom. In principle, simulation of the photoelectron spectrum requires considering all the vibrational degrees of freedom because some vibrational modes may play minor roles in excited state dynamics and yet influence the Franck–Condon factor upon photoionization. The stimulation also requires calculations of unbound electronic states in the ionization continua, which is not a trivial task for theorists. Werner *et al.*[29] have used Stieltjes images (or Stieltjes–Chebyshev moment theory) to describe the ionization, which enabled calculations of PKED but not the photoelectron angular distribution (PAD).

In 1998, we performed time-resolved photoelectron spectroscopy of pyrazine by time-resolved photoelectron imaging (TRPEI) for the first time. However, the time resolution of 450 fs was too low to observe the ultrafast decay of S_2 pyrazine; only S_1 decay ($t \approx 20$ ps) after $S_2 - S_1$ internal conversion was observed.[39,40] A similar study by Stert *et al.* using a magnetic bottle photoelectron spectrometer was also unable to observe ultrafast internal conversion in real time.[41] The development of sub-20 fs deep UV laser (described in the next section) was essential for real-time observation of the $S_2 - S_1$ dynamics of pyrazine.

1.3. Sub-20 fs Deep UV Laser for TRPEI of Pyrazine

Femtosecond lasers that generate pulses with durations of ca. 100 fs are well established as standard light sources for ultrafast spectroscopy.[42,43] A typical system employs a 1 kHz Ti: sapphire regenerative amplifier to pump two computer-controlled collinear parametric amplifiers that generate tunable UV pulses. However, as mentioned above, ultrafast internal conversion cannot be observed in real time using such commercial laser systems because a time resolution of the order of 20 fs is required.

Ultrashort laser pulses in the deep UV to vacuum UV (VUV) region can be generated by optical wave mixing in rare gases that have no photoabsorption in this region. Since a gas medium has a small nonlinear susceptibility, a previous study employed a hollow fiber waveguide filled with a rare gas to obtain a sufficiently long interaction length between the laser pulses and the gas medium.[44,45] However, this method requires excellent laser pointing stabilities and careful coupling of the beams into a narrow fiber channel. In an alternative approach, we used filamentation propagation of intense laser pulses through rare gases.[46] In filamentation propagation, an intense laser pulse travels through a medium, maintaining a small beam diameter for a considerably greater length than the confocal parameter.[47] It is caused by self-focusing of the laser pulse due to the optical Kerr effect and simultaneous diffraction by the plasma created by ionization of the medium. Since filamentation four-wave mixing does not require a fragile hollow fiber, it greatly simplifies the experimental apparatus.

Figure 5 shows a schematic diagram of our filamentation light source and photoelectron imaging spectrometer. The system uses a cryogenically

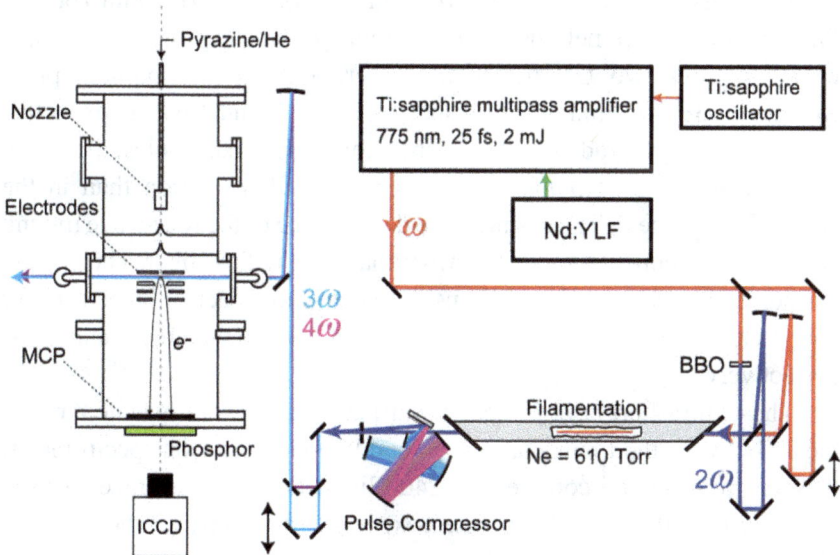

Fig. 5. (Color online) Schematic diagram of filamentation four-wave mixing to generate sub-20 fs deep UV pulses and our photoelectron imaging setup. The red line shows a Gaussian distribution with a FWHM of 22 fs. The cross-correlation between 3ω and 4ω is 22 fs. Reproduced with permission from Ref. 33, copyright (2010) by American Institute of Physics.

cooled Ti:sapphire linear amplifier to generate 25 fs pulses (pulse energy: 2.0 mJ; wavelength: 775 nm; pulse repetition rate: 1 kHz). The fundamental beam (ω) is split into two with an intensity ratio of 7:3 and the high-energy pulse is converted to the second harmonic (2ω) (pulse energy: 0.5 mJ; pulse duration: 30 fs) in a β-barium borate crystal. The second harmonic pulse and the low-energy fundamental pulse (pulse energy: 0.5 mJ) are gently focused into a cell filled with Ne (0.1 MPa). When these two pulses overlap temporally and spatially in the cell, a ~15 cm long bright orange filament appears. The peak powers of these two laser pulses are lower than the critical power for self-focusing in Ne (160 and 40 GW at 775 and 388 nm, respectively).[48] Thus, filamentation propagation is caused by the concerted interactions of multiple laser pulses with the gas medium. Filamentation induces "intensity clamping" and "mode filtering" effects that impart the output laser pulse with a stable energy and an excellent spatial mode.[47] Four-wave mixing produces different harmonics in a single cell by cascaded nonlinear processes: mixing of ω and 2ω simultaneously creates 3ω (264 nm), 4ω (198 nm), and 5ω (157 nm) pulses. This feature is extremely useful for ultrafast photoelectron spectroscopy, which requires two UV pulses for experiments. The maximum pulse energies of 3ω, 4ω, and 5ω are respectively 16, 4, and 1 μJ. As expected, the pulse energy gradually diminishes for higher harmonics; however, the intensity ratio of $5\omega/4\omega/3\omega$ is higher for filamentation than in the hollow fiber process. The gas pressure dependence of the four-wave mixing efficiency in filamentation differs from that for a hollow fiber. The 3ω and 4ω pulses are separated and focused into a photoelectron spectrometer with concave mirrors. The pulse durations of 3ω and 4ω are 14 and 17 fs, respectively.

These ultrashort laser pulses are chirped when they propagate through the air between the filamentation cell and the photoelectron spectrometer; thus, a grating-based compressor (2400 lines/mm, 250 nm braze) is used to compensate the chirp. In our latest experimental setup (not shown here), all the optical paths between the filamentation cell and the photoelectron spectrometer are under vacuum. Consequently, a compressor is not required and deep UV and VUV pulses from the filamentation cell are directly used for measurements.

1.4. Time-Resolved Photoelectron Imaging

TRPEI is photoelectron spectroscopy that employs an ultrafast pump-probe method and two-dimensional position-sensitive detection of electrons. When gas-phase molecules are ionized by a laser pulse in the vacuum chamber, an expanding distribution of photoelectrons is created. The photoelectron imaging (PEI) technique accelerates this distribution in a static electric field and projects it onto a two-dimensional position-sensitive detector that consists of a microchannel plate (MCP), a phosphor screen, and a digital camera (Fig. 5).[49,50] The MCP is a 70 mm diameter circular plate that has millions of 10 μm diameter microchannels over its entire area; the total open area of the microchannels on the MCP surface is ca. 60%. When an electron enters one of the microchannels, an avalanche of secondary electrons occurs and an amplified electron pulse is emitted from the other side of the microchannel. This pulse excites a phosphor screen, visualizing the arrival position of the photoelectrons. The image of the light spot on the phosphor screen is recorded by a charge-coupled device (CCD) sensor or a complementary metal–oxide–semiconductor (CMOS) image sensor. The acceleration electric field is designed such that the arrival positions of electrons depend only on their velocity vectors and not on the spatial positions of ionization. This method thus produces an image of the distribution in k-space (i.e., momentum space).

The simplest electrode design for accelerating electrons employs only three plates,[49,50] and is similar to a Wiley–McLaren time-of-flight mass spectrometer. Using more electrodes improves the spectrometer performance, as demonstrated by Lin *et al.* for ion imaging.[51] The design of our electrodes is shown in Fig. 6.[52] A large square hole in electrode 4 allows the propagation of laser beams or He(I) radiation. The ionization point is indicated by a cross ($\times$). We designed electrodes 1–3 to reduce background photoemission due to stray light. Electrode 3 is a repeller plate, but it has a large hole in the center to reduce background photoemission. To flatten the equipotential around the hole in electrode 3, electrode 2 is held at the same voltage as electrode 3. Electrode 1 is used to prevent the ground potential penetrating the acceleration electric field, while a high-transmission (90%) mesh minimizes its cross-section and consequently the background photoemission from electrode 1. Electrode 1 is at a slightly

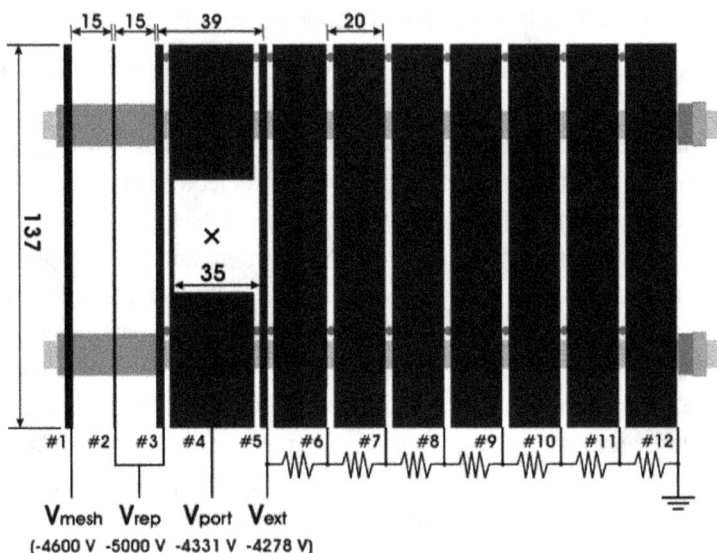

Fig. 6. Cross-sectional view of our electrostatic lens system (all dimensions in millimeters). A molecular beam is introduced from the left. The ionization point is indicated by the cross ($\times$).

higher potential than electrodes 2 and 3 to prevent photoelectrons emitted from electrode 1 due to stray light being transmitted toward the detector. Electrode 1 has a 6 mm diameter hole in its center to allow the molecular beam to propagate parallel to the axis of the electrode stack. We computed electron trajectories and found that the velocity resolution improves when even more electrodes are used; however, the velocity resolution saturates in practice. The design shown in Fig. 5 provides $\Delta v/v < 0.04\,\%$ for a focused laser beam.

Even if velocity map imaging electrodes focus the trajectories of electrons with the same velocity, the light spot on the phosphor screen will be considerably larger in diameter than the microchannel pore diameter ($10\,\mu$m), thereby causing blurring of the photoelectron image. Thus, the center of gravity should be calculated to recover the ultimate spatial resolution provided by the velocity map imaging electrodes. The center of gravity can be recorded only when the brightness and/or the area of the light spot exceeds a preset threshold for two-dimensional electron counting. This ensures a uniform detection sensitivity over the detector area and highly reliable experimental results.[52-54] To perform center-of-gravity calculations and electron counting for each laser shot, the frame rate of the camera must be comparable with or higher than the repetition rate (1 kHz) of the femtosecond laser. We thus constructed a 1 kHz camera using a CMOS image sensor and a field programmable gate array circuit for real-time image processing. A CMOS sensor has a much faster readout than a CCD sensor, although it has a considerably lower sensitivity than a CCD sensor at the present time. We thus use an image intensifier and booster to improve the sensitivity of our camera system; more details are described in the original paper.[53] The 1 kHz CMOS camera enables highly accurate two-dimensional electron counting to record photoelectron images.

The observed photoelectron image corresponds to a two-dimensional projection of a three-dimensional photoelectron velocity distribution; therefore, the key step in analyzing the PAD is to calculate a slice through a three-dimensional velocity distribution from the observed image. The inverse Abel transform is a mathematical inversion method that does not require fitting, whereas the pBaseX method involves least-squares fitting of the data with a basis set.[55] These two methods generally provide comparable results. However, when the spatial resolution increases, an image contains many dots and tends to become rough. The pBaseX method is more useful than the inverse Abel transform for analyzing such experimental data.

1.4.1. *TRPEI of Ultrafast S_2–S_1 internal conversion in pyrazine*

Figure 7 summarizes the experimental results originally published in two different papers.[33,56] Figure 7(b) shows the pump-probe photoelectron signal obtained for pyrazine-h4 using 264 nm pump and 198 nm probe pulses. As shown in Fig. 4, the spectra of our pump and probe pulses overlap the $S_2 - S_0$ and $S_3 - S_0$ bands, respectively. Consequently, for a positive time delay, the 264 nm pulse excites ground-state molecules to S_2 and the 198 nm pulse ionizes them. For a negative time delay, the roles of the 198 nm and 264 nm pulses are exchanged and molecules are ionized from S_3. The signal at a positive time delay rapidly decays in less than 100 fs and exhibits a plateau; this plateau has a finite lifetime of 22 ps for pyrazine-h4.[39,41] Furthermore, the plateau region exhibits oscillatory features due to vibrational quantum beats. The Fourier transform of the oscillation

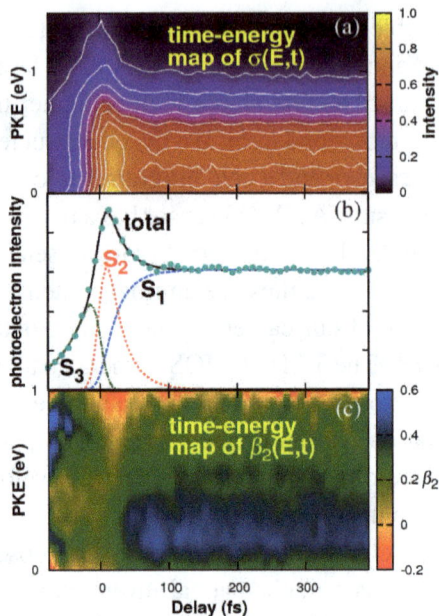

Fig. 7. (Color online) (a) Time-evolution of PKED, $\sigma(E, t)$. (b) Temporal profiles of total photoelectron signals in $(1 + 1')$ TRPEI of pyrazine-h4. The observed data are well explained by three components: single-exponential decay of S_2 (red), corresponding increase in S_1 (blue) at a positive time delay, and single-exponential decay of S_3 (green) at a negative time delay. The fitting result is shown by the solid line. (c) Time-evolution of photoelectron angular anisotropy parameter $\beta_2(E, t)$.

$(t > 50\,\text{fs})$ exhibits a frequency component of $560 \pm 40\,\text{cm}^{-1}$, which agrees with the vibrational frequency of Q_{6a} in $S_1\,(583\,\text{cm}^{-1})$. Similarly, pyrazine-d4 exhibits a Fourier component of $550 \pm 40\,\text{cm}^{-1}$, which provides further support for the assignment to $Q_{6a}\,(\nu_{6a}(S_1) = 564\,\text{cm}^{-1}$ for pyrazine-d4).[57] In the negative time range, the signal also diminishes very rapidly within $100\,\text{fs}$ (toward the $-\infty$ direction). The observed profile can thus be explained by three components: the decay of optically excited S_2, the corresponding growth of S_1 populated by internal conversion from S_2, and the decay of S_3. By least-squares fitting, the $S_2 \rightarrow S_1$ internal conversion time constants are estimated to be $23 \pm 4\,\text{fs}$ for pyrazine-h4 and $20 \pm 2\,\text{fs}$ for pyrazine-d4. The time constants for S_3 decay are $43 \pm 3\,\text{fs}$ for pyrazine-h4 and $44 \pm 3\,\text{fs}$ for pyrazine-d4.

The time-dependent photoionization differential cross-section in $(1 + 1')$ resonance-enhanced multiphoton ionization (REMPI) with linearly polarized pump and probe light, where the polarizations are parallel to each other, is expressed by

$$I(E, \theta, t) = \frac{\sigma(E, t)}{4\pi}\{1 + \beta_2(E, t)P_2(\cos \theta) + \beta_4(E, t)P_4(\cos \theta)\}$$

$$= \sum_{i=1,2,3} \frac{\sigma^{(i)}(E, t)}{4\pi}\{1 + \beta_2^{(i)}(E)P_2(\cos \theta) + \beta_4^{(i)}(E)P_4(\cos \theta)\},$$

$$(1)$$

where E is the photoelectron kinetic energy, θ is the electron ejection angle relative to the laser polarization direction, and t is the pump-probe time delay. $P_n(x)$ is the nth-order Legendre polynomial and $i = 1, 2$, and 3 correspond to the S_1, S_2, and S_3 components, respectively. $\sigma^{(i)}(E, t)$ describes a PKED for ionization from each electronic state as a function of time, which is dictated by the energetics of the ionization process including the Franck–Condon envelope on ionization. For symmetry reasons, interference is not expected for photoionization from S_1 and S_2. Integrating Eq. (1) over the scattering angle gives a time-dependent PKED, i.e., $\sigma(E, t) = \Sigma_{i=1,2,3}\sigma^{(i)}(E, t)$, as shown in Fig. 7(b).

$I(E, t)$ does not exhibit any marked change on $S_2 \rightarrow S_1$ internal conversion. This indicates that photoionization occurs predominantly as $D_0(n^{-1}) \leftarrow S_1(n, \pi^*)$ and $D_1(\pi^{-1}) \leftarrow S_2(\pi, \pi^*)$, in accordance with the one-electron model of photoionization with a frozen core. Since the energy gaps between D_1 and D_0 (0.88 eV) and between S_2 and S_1 (0.86 eV) are very similar, the PKEDs for these two processes are almost the same.

Figure 7(c) shows a time-energy map of $\beta_2(E, t)$. The positive (blue–green) and negative (red) values correspond to preferential ejection of an electron parallel and perpendicular to the probe laser polarization [see Eq. (1)]. The energy dependence of β_2, a colored stripe at each time delay in Fig. 7(b), is a fingerprint of the electronic character. The time-energy map clearly shows that there are three different components, one at a negative time delay and two at a positive time delay, which agrees with the analysis of $\sigma(E, t)$. The most distinctive feature is the sudden change in the color at ca. 30 fs, which is attributed to ultrafast $S_2 \rightarrow S_1$ internal conversion.

$\beta_2(E, t)$ does not change after 30 fs, indicating that the (n, π^*) electronic character remains; no restoration of the (π, π^*) character is identified. This lack of recurrence is possibly related to the photoexcitation energy; we excited pyrazine near the S_2 origin with the pump laser spectrum indicated in Fig. 4 to prepare a wave packet with a small vibrational excess energy in S_2. Consequently, if the vibrational energy flows into various modes in S_1, the wave packet has no chance to return to the Franck–Condon region in S_2. Photoexcitation at a shorter wavelength to reach higher vibronic levels in S_2 may enable restoration of the (π, π^*) character.

Figure 8 compares the observed PKED with the results of on-the-fly molecular dynamics simulations, which calculate classical mechanical nuclear motions using the forces obtained by time-dependent density

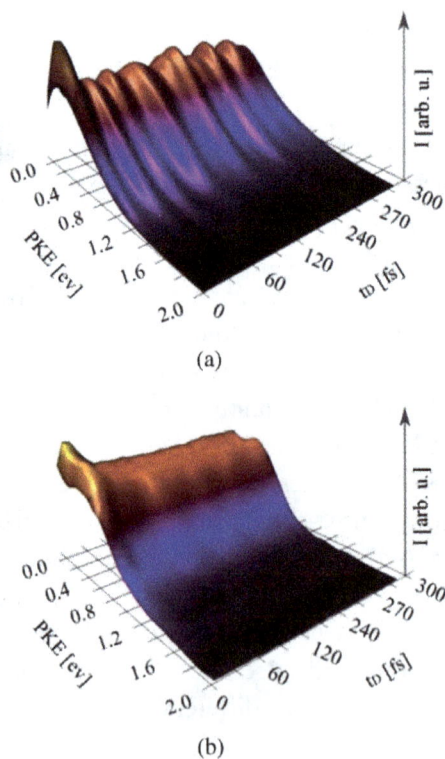

(a)

(b)

Fig. 8.   Time-dependent PKED in TRPES of pyrazine: (a) theoretical simulation using on-the-fly molecular dynamics and Stieltjes imaging and (b) experimental result.

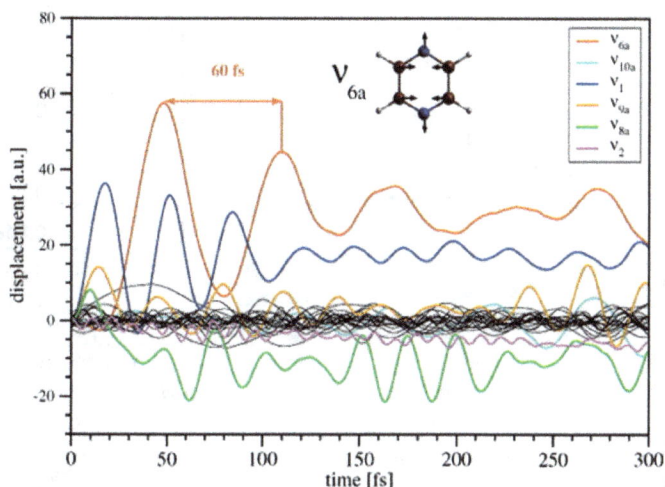

Fig. 9. Time-dependent normal mode displacements obtained by projection onto the equilibrium ground state normal coordinates of pyrazine averaged over 60 trajectories. The displacement of the dominant normal mode Q_{6a} exhibits a periodicity of ~60 fs. Reproduced with permission from Ref. 29, copyright (2010) by American Institute of Physics.

functional theory. Although the oscillatory feature in the plateau is enhanced in the molecular dynamics simulation, the magnitudes of the oscillations are rather small in both experiment and calculations. This is attributed to destructive interference among the classical oscillations of different normal modes. Figure 9 shows the calculated time-dependent displacement along the normal modes of pyrazine in the ground state.

Close examination of the experimental PKED reveals that the vibrational quantum beat of Q_{6a} has slightly different amplitudes at different PKE, as shown in Fig. 10. The Fourier transforms of these oscillatory components are shown in Figs. 10(c) and 10(d). The strongest beat was observed at a PKE of 0.64 eV. Pyrazine-d4 exhibits smaller beat amplitudes. The time profiles extracted for different PKEs from a series of photoelectron images have noise, which reduces the quality of the Fourier transforms. If the observed beat is attributed to motion on S_2, the initial phase of the quantum beat should be 0 or π.[58] On the other hand, the observed quantum beats of photoelectron signals in Figs. 8(b) or 10, exhibit nonzero initial phases when the observed photoelectron signals are fitted with a functional form of $A(E) + B(E) \cos[\omega t + \varphi(E)]$; the initial phase $\varphi(E)$ is determined

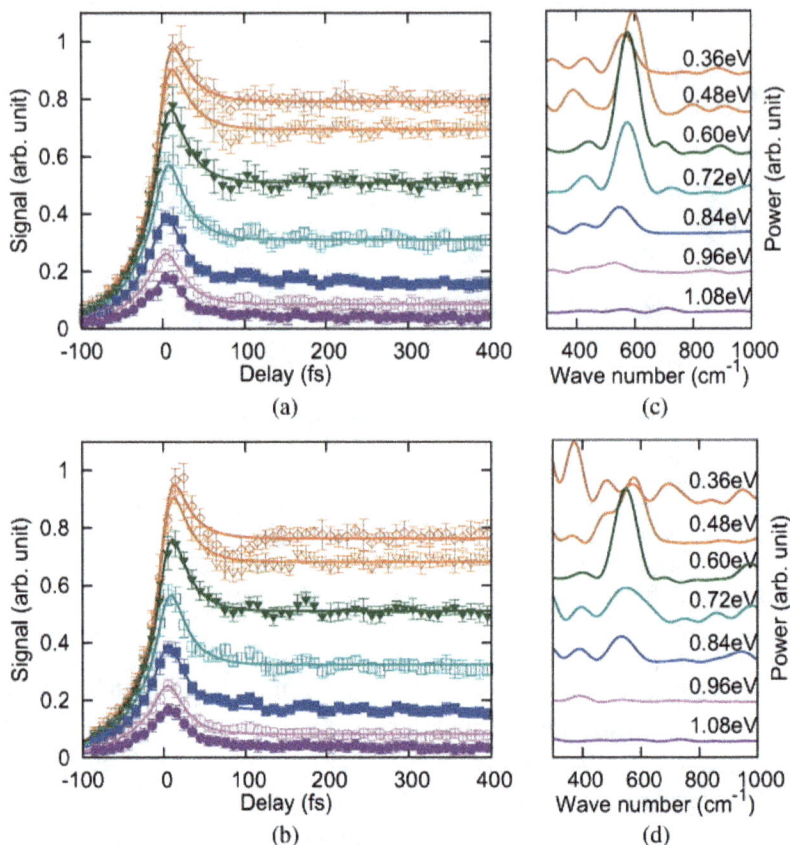

Fig. 10.   Time evolution of photoelectron intensities at selected PKE subsections of (a) pyrazine-h4 and (b) pyrazine-d4. The oscillatory features are due to vibrational wave-packet motion along Q_{6a}. The solid lines show the results of global fitting. Fourier power spectra of oscillatory components of (c) pyrazine-h4 and (d) pyrazine-d4 photoelectron signal intensities after a delay time of 70 fs are also shown. Reproduced with permission from Ref. 33, copyright (2010) by American Institute of Physics.

to be ca $0.35(\pm 0.1) \times \pi$ radians in the range $0.24 < E < 0.92$ eV. The nonzero initial phase confirms that the observed quantum beat is not due to vibrational motion on the S_2 potential energy surface.

We performed global fitting of the spectrum for all energies and all time delays by using the Levenberg–Marquardt algorithm to extract photoelectron spectra in ionization from S_1, S_2, and S_3. The signal is expressed in terms of the following three time-dependent functions:

$$I(E_i, t) = \{C_{1,i}(1 - \exp(-t/\tau_2)) + C_{2,i}\exp(-t/\tau_2)$$
$$+ C_{3,i}\exp(-(-t)/\tau_3)\} \otimes g(t) \quad (2)$$

where E_i is the kinetic energy bin and $g(t)$ is a Gaussian cross correlation of 22 fs (FWHM). $C_{1,i}$, $C_{2,i}$, and $C_{3,i}$, respectively correspond to the amplitudes of the S_1, S_2, and S_3 contributions at a PKE of E_i. In our fitting, only the amplitudes $C_{j,k}$ are allowed to vary across the spectrum. The best fit is obtained for time constants of $\tau_2 = 22.4 \pm 2.0$ fs and $\tau_3 = 40.6 \pm 1.5$ fs for pyrazine-h4 and $\tau_2 = 21.8 \pm 1.7$ fs and $\tau_3 = 42.7 \pm 1.7$ fs for pyrazine-d4. The solid lines in Fig. 10 represent fitted curves at different photoelectron energies. The global fitting reproduces the observed signal very well, except that it does not account for the oscillatory feature. A series of $C_{j,k}$ for a given j corresponds to a photoelectron spectrum in photoionization from S_j to D_j. The obtained $C_{j,k}$ coefficients are shown in Fig. 11. The dashed lines in Fig. 11 show the maximum electron energies (E_i^{max}) for ionization to each D_j state given by $E_i^{max} = \hbar\omega_{pump} + \hbar\omega_{probe} - \varepsilon_i$, where ω_{pump} and ω_{probe} are respectively the optical angular frequencies of the pump and probe pulses and ε_i is the adiabatic ionization energy to D_i. Figure 11 indicates that ionization from both S_2 and S_1 occurs at least in part to D_0. The independent electron approximation predicts that S_1 will be predominantly ionized to D_0. Since the S_1 state created by internal conversion from S_2 has a vibrational energy as large as 1 eV, the maximum PKED in $D_0 \leftarrow S_1$ ionization is shifted by 1 eV, if S_1 and D_0 have identical potential energy surfaces. In reality, since the potentials differ slightly, the maximum PKED appears at a lower PKE and hence a higher vibrational energy in D_0. The appearance of $D_0(n^{-1}) \leftarrow S_2(\pi\pi^*)$ ionization indicates the breakdown of the independent electron approximation, as discussed in more detail in the next section.

1.4.2. *Analysis of PAD*

The PAD is the angular dependence of the probability density (squared amplitude of the wave function) of a photoelectron far from the ion core. The PAD observed by TRPEI is the distribution measured in the laboratory frame; it is thus more unambiguously termed the laboratory-frame PAD

(a)

(b)

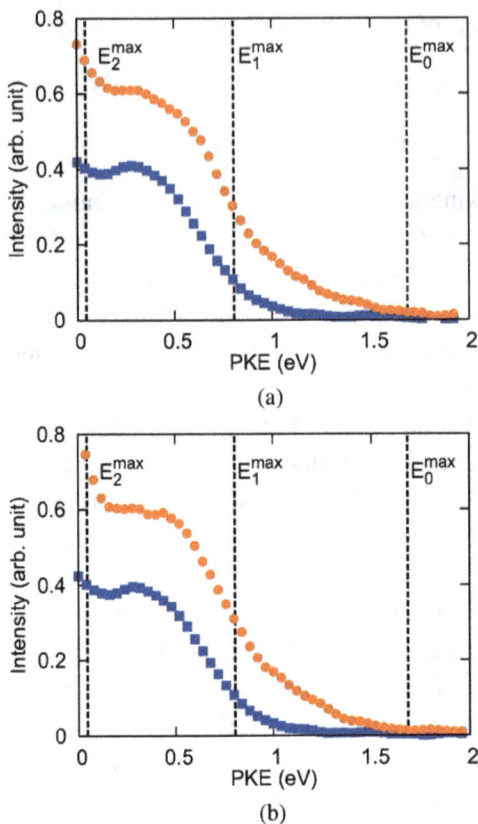

Fig. 11.   Fitted coefficients for intensities of S_1 (squares) and S_2 (circles) components for (a) pyrazine-h4 and (b) pyrazine-d4. Each dashed line (E_i^{max}) corresponds to the maximum possible energy for ionization to the vibrational ground state of each ionic state, D_i. The enhanced intensities for 0–0.2 eV may be ascribed to background electrons and neglected in the discussion. Reproduced with permission from Ref. 33, copyright (2010) by American Institute of Physics.

(LF-PAD). On the other hand, the PAD determined with respect to the symmetry axis of a molecule is termed the molecular-frame PAD (MF-PAD). In our study, we experimentally observe the LF-PAD and interpret it by comparing with simulations of the LF-PAD based on the MF-PAD computed using some approximations. The asymptotic photoelectron wave function in the molecular frame is usually expressed by a linear combination of partial waves (which are spherical harmonics to simplify computations, even though the molecules are not spherically symmetric).

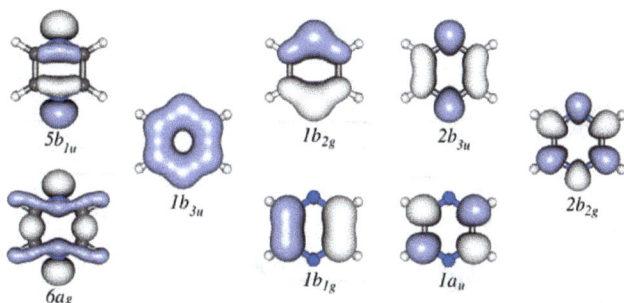

Fig. 12. Outer valence and π^* orbitals of pyrazine obtained by Hartree–Fock calculations using a minimal basis set (HF/STO-3G).

(Unbound wave functions that satisfy the symmetry requirement of a given point group of a molecule can be constructed from a linear combination of spherical harmonics; such wave functions are termed eigenchannel wave functions.) The coefficients of these partial waves are the transition dipole moments from the ionized orbital to each partial wave. In the independent electron approximation, ionizations from S_2 and S_1 to D_1 and D_0, respectively, have the same ionized orbital, $\pi^*(2b_{3u})$ (Fig. 12). Consequently, the transition dipole moments and hence the PADs are expected to be the same for the two processes, whereas the observed anisotropy parameters differ for ionization from S_2 and S_1. In this section, we consider the origin of this difference. For this purpose, we simulated PADs by the first-order configuration interaction (FOCI)[59] method and by using the continuum multiple scattering $X\alpha$ (CMSX-α) approximation.[60]

Notice that ionization processes creating PKE < 0.8 eV are mainly $D_1 \leftarrow S_2$ ionization and $D_0 \leftarrow S_1$, while PKE > 0.8 eV is created by $D_0 \leftarrow S_2$ and $D_0 \leftarrow S_1$. Thus, the photoelectron angular anisotropy in these two regions should be considered separately. We discuss the latter region first. Ionization from S_2 in that region is solely due to $D_0(n^{-1}) \leftarrow S_2(\pi\pi^*)$. This process is forbidden for the main electron configurations of D_0 and S_2; therefore, the occurrence of this ionization process indicates that D_0 and S_2 consist of multiple electron configurations. The D_0 configurations that can be created by one-photon ionization from S_2 are those obtained by removing one electron from an orbital (φ) of the S_2 configuration, i.e.,

$$\Psi(S_2) = \Psi(D_0) \times \varphi, \qquad (3)$$

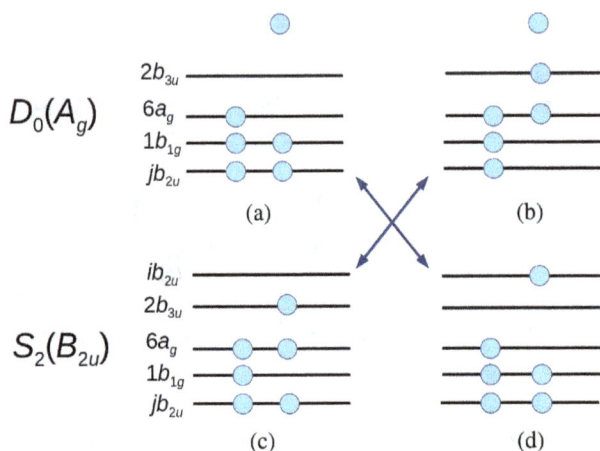

Fig. 13.   Electronic configurations of D_0 and S_2. The leading configurations of D_0 and S_2 are shown in (a) and (c), respectively. By two-electron excitation, configurations (b) and (d) are obtained from configurations (a) and (c), respectively. Arrows indicate the allowed transition by one-photon ionization. Filled circles represent electrons. Isolated electrons at the top of panels (a) and (b) represent photoelectrons.

where $\Psi(S_2)$ and $\Psi(D_0)$, respectively denote the electron configurations of S_2 and D_0. Because $\Psi(D_0) \times \varphi$ should have the same symmetry species as $S_2[\Gamma(S_2) = \Gamma(D_0) \times \Gamma(\varphi)]$, φ must be the b_{2u} orbital as given by the direct product $A_g(D_0) \times B_{2u}(S_2)$. However, no b_{2u} orbital exists among the outer valence and π^* orbitals, which implies that $D_0 \leftarrow S_2$ cannot be well described by typical valence complete active space self-consistent field (CASSCF) wave functions.[61] To make the calculations tractable, we focused on configurations that are doubly excited with respect to the main configuration. Figure 13 shows examples of such configurations. The configurations shown in Figs. 13(b) and 13(d) can be obtained by two-electron excitations from the D_0 and S_2 main configurations shown in Figs. 13(a) and 13(c), respectively. Ionization from the configuration in Fig. 13(d) to that in Fig. 13(a) and from the configuration in Fig. 13(c) to Fig. 13(b) is possible. Including these configurations, FOCI calculations account for all one-electron excitations from the complete active space of eight orbitals [n, π, π^* (Fig. 12)].

Figures 14(a)–14(c) show polar plots of the observed PADs at a PKE of 0.9 eV for different pump-probe delay times. The PAD exhibits enhanced

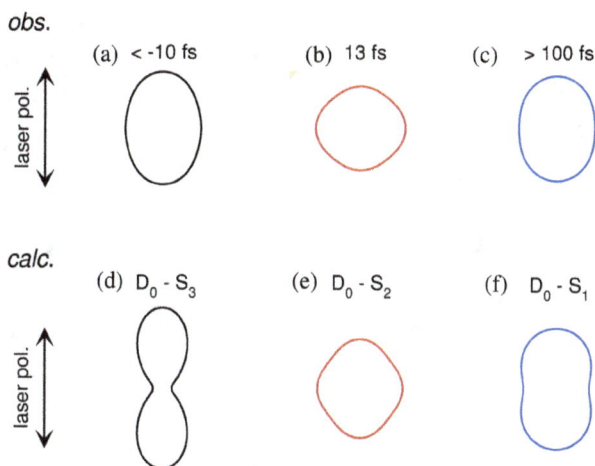

Fig. 14.   Polar plots of observed and calculated PADs at 0.9 eV. The observed PADs (a) and (c) were averaged from − 47 to − 10 fs and from 100 to 393 fs, respectively. (b) Observed PAD at 13 fs. (d), (e), and (f) show calculated PADs for ionizations from S_3, S_2, and S_1 to D_0, respectively.

intensities along the laser polarization for negative and positive time delays, while it is almost isotropic at time zero. As mentioned above, all these observed signals are due to ionization to D_0. The calculated PADs for $D_0 − S_3$, $D_0 − S_2$, and $D_0 − S_1$ (shown in Figs. 14(d)–14(f), respectively) reproduce the observed features rather well.

Figure 15 compares the observed and calculated PADs at a PKE of 0.5 eV. The calculated PADs are generally in reasonable agreement with the observed ones; however, calculations predict a smaller change in the photoelectron angular anisotropy on $S_2 − S_1$ internal conversion than that observed. There are three possible reasons for this discrepancy. First, ionization from S_2 to both D_0 and D_1 can create this low photoelectron kinetic energy, where $D_0 − S_2$ provides a smaller $\beta_2(E)$ than $D_1 − S_2$. Thus, the experimental result that involves the $D_0 − S_2$ photoelectron signals may exhibit a smaller $\beta_2(E)$ than the calculation that assumes only $D_1 − S_2$ (Figs. 15(b) and 15(c)). Second, although both $D_1 − S_2$ and $D_0 − S_1$ processes induce ejection of the same $\pi^*(2b_{3u})$ electron, they create cations in different electronic states. FOCI/CMSX-α calculations may not be sufficiently accurate for electron-ion interaction potential at short electron-ion distances. A third possible reason for the difference in

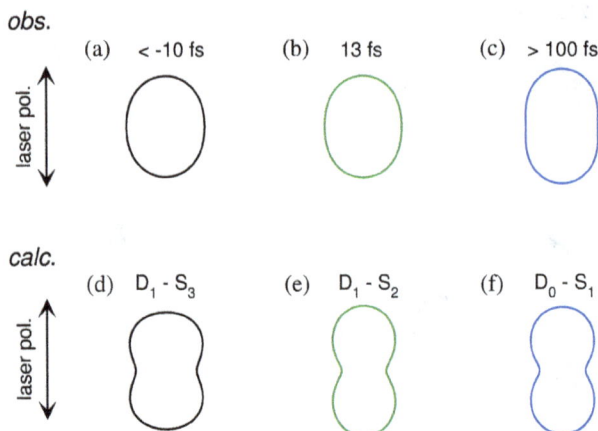

Fig. 15. Polar plots of observed and calculated PADs at 0.5 eV. The observed PADs (a) and (c) were averaged from -47 to -10 fs and from 100 fs to 393 fs (b) observed PAD at 13 fs. (d) and (e) show calculated PADs for ionizations from S_3 and S_2, respectively, to D_1 while (f) from S_1 to D_0.

β_2 is a difference in the molecular geometry for S_2 and S_1 at the instant of ionization; S_2 is mainly ionized from the Franck–Condon state, which is similar in structure to the equilibrium geometry in S_0, while S_1 is ionized from the vibrationally excited states with different geometries from S_0 along several normal coordinates. Our FOCI/CMSX-α calculations were performed for the equilibrium geometry in S_0; they did not account for the structural difference between S_2 and S_1.

The difference between Figs. 14(f) and 15(f) indicates that the photoelectron anisotropy parameters depend on the PKE. This is understood by considering the energy-dependent phase factors (Coulomb phases) of partial waves. According to FOCI/CMSX-α calculations for pyrazine, the relevant partial waves for $D_0 \leftarrow S_1$ are the d and g waves; the s-wave contribution is small. Thus, the reduction in $\beta_2(E)$ with increasing PKE on ionization from S_1 is attributed to interference caused by the energy-dependent Coulomb phases of the d and g waves. $\beta_2(E)$ of $D_0 \leftarrow S_2$ also decreases with increasing PKE. This reduction is enhanced by shape resonance in the kb_{3g} continuum at 3.5 eV.[60,62]

In summary, FOCI/CMSX-α calculations indicate that the negative β_2 observed for $D_0 - S_2$ ionization is caused by shape resonance. Time-energy mapping enables unambiguous identification of the time-dependent electronic character and nonadiabatic transitions, even if there are influences from shape resonances. More accurate calculations will be helpful for explaining the difference in β_2 for $D_1 \leftarrow S_2$ and $D_0 \leftarrow S_1$. Recently, Arasaki *et al.* have theoretically simulated TRPEI of nonadiabatic dynamics of NO_2 in the molecular frame using quantum wave packet calculations for excited-state dynamics and Schwinger variational calculations for photoionization dynamics.[63] Their calculations demonstrated the usefulness of time and energy-resolved PADs for analyzing nonadiabatic dynamics. Similar theoretical calculations on pyrazine would deepen our understanding of the nonadiabatic dynamics of this benchmark system.

1.5. Conical Intersections in Cation and Rydberg States of Pyrazine

Similar to $S_2(\pi\pi^*)$ and $S_1(n\pi^*)$, the $D_1(\pi^{-1})$ and $D_0(n^{-1})$ potential energy surfaces of pyrazine have a conical intersection.[64] This intersection in the cation raises some interesting questions. First, if ultrafast internal conversion occurs from $D_1(\pi^{-1})$, lifetime broadening should occur in the $D_1(\pi^{-1}) \leftarrow S_0$ spectrum. Since photoelectron spectra were previously measured for pyrazine vapor, lifetime broadening was not well discriminated from the influences of the rotational envelopes and vibrational hot bands. For unambiguous discussion of lifetime broadening, the photoelectron spectrum should be measured at ultralow temperatures. Second, if ultrafast internal conversion occurs in the cation, similar processes may occur in the Rydberg states because the Rydberg states consist of the same ion core as the cation and a loosely bound Rydberg electron. The question then arises as to whether it is possible to observe the zero kinetic energy photoelectron or pulsed field ionization photoelectron (PFI-PE) spectrum for the $D_1(\pi^{-1})$ state of pyrazine. PFI-PE spectroscopy creates Rydberg states with extremely high principal quantum numbers and high angular momentum quantum numbers and field ionizes them by a pulsed electric field. By scanning the laser wavelength and monitoring the yield of electrons or ions on field ionization, PFI-PE spectroscopy measures an action spectrum that is similar to a conventional photoelectron spectrum. In the case of pyrazine, is it still possible to observe the PFI-PE spectrum even when ultrafast internal conversion occurs in the ion cores of the Rydberg states?

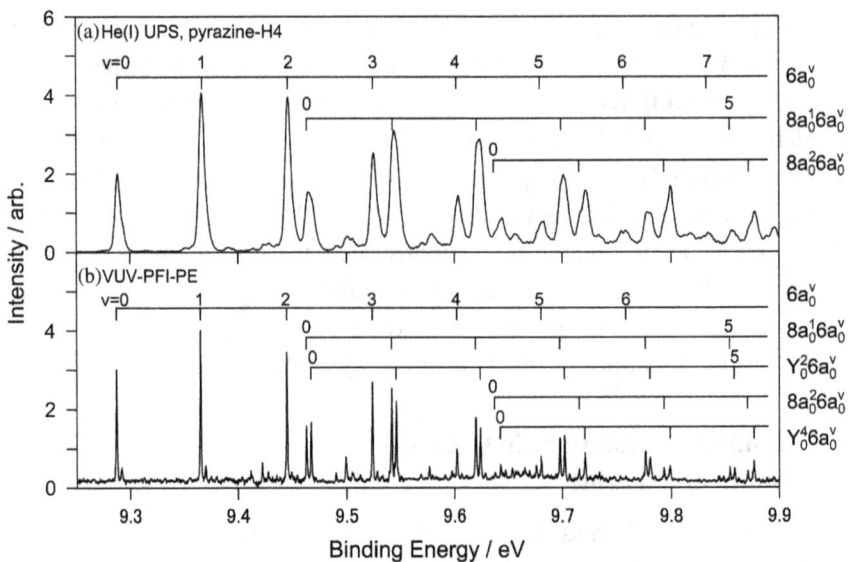

Fig. 16.   (a) Expanded view of He(I) photoelectron spectrum of pyrazine with vibrational assignments. (b) VUV-PFI-PE spectra in the $D_0(n^{-1}) \leftarrow S_0$ region. Reproduced with permission from Ref. 65, copyright (2008) by American Chemical Society.

Figure 16(a) shows the He(I) photoelectron spectrum of jet-cooled pyrazine measured using a He discharge lamp and a hemispherical electron energy analyzer and Fig. 16(b) shows the corresponding region of the PFI-PE spectrum.[65] While both these spectra show one-photon photoionization from the ground electronic state, the former shows direct photoionization, whereas the latter shows resonant excitation to Rydberg states that are energetically almost degenerate with the cation states. Due to the structural change caused by the removal of a valence electron, these spectra exhibit rich vibrational structures that are in remarkable agreement with each other. Close examination reveals that the vibrational temperature is lower in PFI-PE because it employs pulsed expansion of the gas sample to achieve a low vibrational temperature.[65] In contrast, He(I) photoelectron spectroscopy uses a continuous gas jet. The He(I) photoelectron spectrometer has resolutions of 5.5 meV and 9 meV for pyrazine and fully deuterated pyrazine, respectively, while that of PFI-PE is 1.5 cm^{-1} (0.2 meV).

We examine the $D_1(\pi^{-1})$ region in Fig. 17, which compares the He(I) photoelectron spectra of pyrazine vapor previously reported,[66] jet-cooled

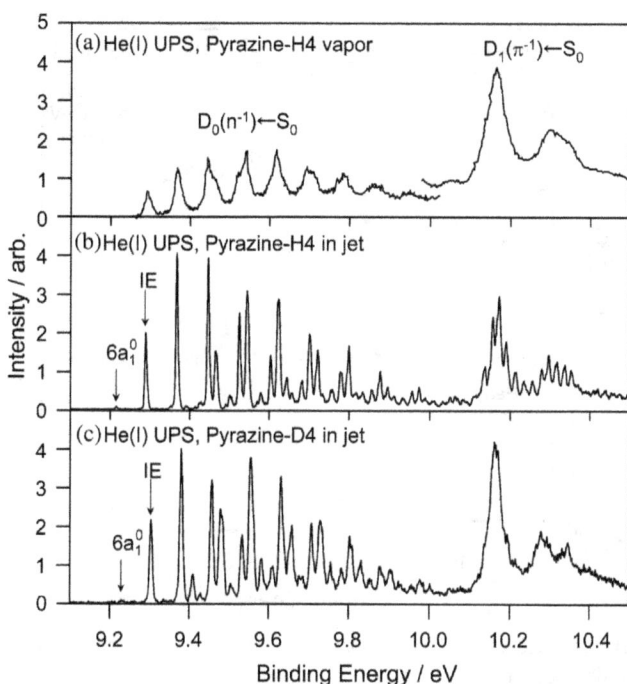

Fig. 17. (a) He(I) photoelectron spectrum at room temperature reproduced from Ref. 65 with energy recalibration by 83 meV. (b) He(I) UPS of pyrazine in a supersonic jet. The spectral resolution is 5.5 meV. (c) He(I) photoelectron spectrum of fully deuterated pyrazine in a supersonic jet. The spectral resolution is 9 meV. Reproduced with permission from Ref. 65, copyright (2008) by American Chemical Society.

pyrazine and fully deuterated pyrazine. Comparison of the photoelectron spectrum of pyrazine vapor (Fig. 17(a)) with our spectrum of a jet-cooled sample (Fig. 17(b)) clearly reveals that the former suffers from instrumental limitations. Our spectra are considerably sharper than the previously obtained spectrum due to supersonic jet cooling of the sample and a higher spectral resolution. The difference in the spectral features in the $D_1(\pi^{-1})$ region is striking: Fig. 17(a) shows only a few broad bands, whereas each of these bands is split into several bands in Fig. 17(b). Interestingly, the same fine splitting is not observed for fully deuterated pyrazine (Fig. 17(c)). Figure 18 presents expanded views of the $D_1(\pi^{-1})$ region in the three spectra measured for jet-cooled samples. The PFI-PE spectrum in Fig. 18(b) contains sharp bands in the $D_1(\pi^{-1})$ region. However, their features are

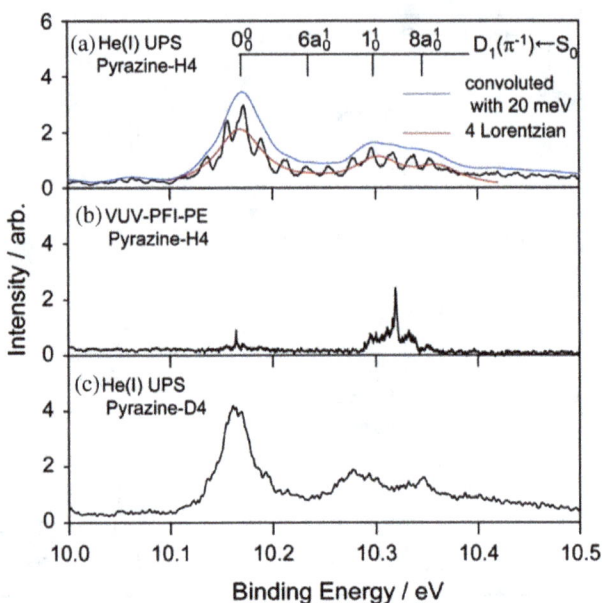

Fig. 18.   (a) Expanded view of He(I) photoelectron spectrum of jet-cooled pyrazine with vibrational assignments in the $D_1(\pi^{-1}) - S_0$ region. Convolution of the observed spectrum with a virtual instrumental resolution of 20 meV erases structures due to fine splitting. The envelope of the spectral feature is reproduced using four Lorentzian functions for the bands indicated in the figure. (b) VUV-PFI-PE spectrum of pyrazine in the $D_1(\pi^{-1}) - S_0$ region. (c) He(I) photoelectron spectrum of fully deuterated pyrazine in a supersonic jet. Reproduced with permission from Ref. 65, copyright (2008) by American Chemical Society.

completely different from those in the He(I) photoelectron spectrum shown in Fig. 18(a). This result demonstrates that it is difficult to observe a PFI-PE spectrum for the $D_1(\pi^{-1})$ state that undergoes ultrafast internal conversion. We conjecture that the internal conversion mediates couplings with dissociative neutral states and/or ionization continua to induce dissociation into neutral fragments and autoionization. From spectral fitting, the lifetimes of the $D_1(\pi^{-1})$ states of pyrazine and fully deuterated pyrazine are estimated to be 12 fs and 15 fs, respectively.

To estimate the location of the conical intersection point, we analyzed the Franck–Condon factors of the $D_0(n^{-1})$ and $D_1(\pi^{-1})$ bands. As seen in Fig. 16, $D_0(n^{-1}) \leftarrow S_0$ exhibits vibrational progressions of 6a and 8a modes. On the other hand, the $D_1(\pi^{-1}) \leftarrow S_0$ spectrum exhibits a strong 0–0 band, indicating that the equilibrium geometry in D_1 is almost

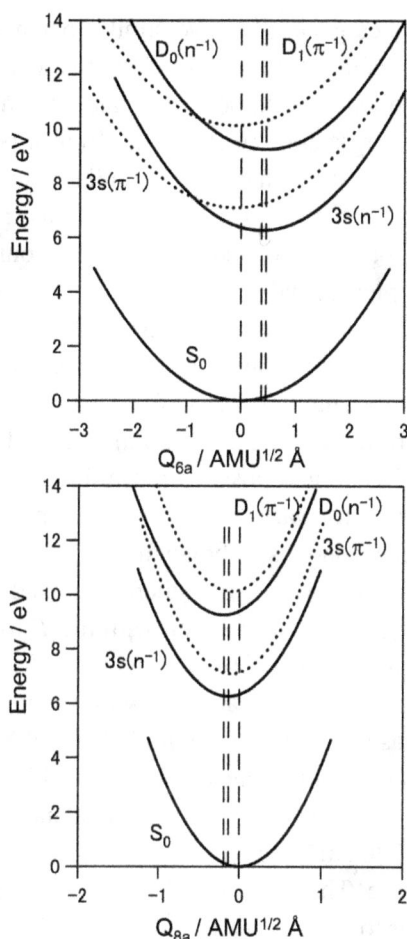

Fig. 19. Harmonic potential curves along with $6a$ and $8a$ normal coordinates for pyrazine determined from spectroscopic data. The equilibrium geometries in the $3s(n^{-1})$ and $D_0(n^{-1})$ states differ significantly from that of the ground state. The equilibrium geometry of the $3s(n^{-1})$ state differs from that of the $D_0(n^{-1})$ state. Reproduced with permission from Ref. 65, copyright (2008) by American Chemical Society.

the same as S_0. Franck–Condon analysis provides the magnitudes of the displacements ΔQ, but not their signs. Therefore, we determined their signs based on the calculated equilibrium geometry of $D_0(n^{-1})$ at the B3LYP/cc-pVTZ level. Figure 19 shows the harmonic potential curves along the $6a$ and $8a$ normal coordinates. Crossings of these potentials are clearly observed for the $6a$ mode.

The Rydberg states generally have similar potential energy surfaces as those of the cation since the Rydberg electrons with high principal and angular momentum quantum numbers penetrate little into the ion core. For the lowest (3s) Rydberg state, the Rydberg electron penetrates relatively deeply into the core, but it still has quite a similar potential energy surface to that of the cation. Our first study of the 3s Rydberg states of pyrazine was performed using a femtosecond laser, whereas our second study was performed using a picosecond laser. These $(2 + 1)$ REMPI spectra of pyrazine via 3s Rydberg states are shown in Fig. 20 along with the spectrum recorded using a nanosecond laser. All three spectra were recorded by scanning the laser wavelength while monitoring the photoionization signal intensity. The spectrum recorded using a femtosecond laser is very broad due to its wide bandwidth and possible power broadening. This spectrum has not been corrected for variation in the laser intensity. The spectrum measured with a picosecond laser shown in Fig. 20(b) exhibits a very clear vibrational feature for the $3s(n^{-1})$ Rydberg state and a broad feature for the $3s(\pi^{-1})$ Rydberg state. Comparison with the $D_0(n^{-1})$ photoelectron spectrum reveals that the $3s(n^{-1})$ state mainly differs in that it exhibits lifetime broadening due to interactions with valence electronic states, whereas $D_0(n^{-1})$ has no decay; the vibronic band of $3s(n^{-1}) \leftarrow S_0$ has a width of $15 \, \mathrm{cm}^{-1}$. Our main interest here is the width of the 0–0 band of $3s(\pi^{-1}) \leftarrow S_0$; it is as large as $390 \, \mathrm{cm}^{-1}$, corresponding to a lifetime of 14 fs. A similar width, $370 \, \mathrm{cm}^{-1}$, is observed for the $3s(\pi^{-1}) \leftarrow S_0$ 0–0 band of deuterated pyrazine. The lifetimes of the $3s(\pi^{-1})$ Rydberg states thus estimated are similar to those of $D_1(\pi^{-1})$. Our study clearly demonstrates that ultrafast internal conversion in the ion core also occurs in the Rydberg states.

Figure 20 also shows the PAD measured for each vibronic bands.[67] The PADs observed for $3s(n^{-1})$ and $3s(\pi^{-1})$ differ greatly, which assists assignment of vibronic bands. PEI is expected to be useful for analyzing complex photoabsorption spectra of higher excited states.

1.6. Toward Sub-30 fs TRPEI in VUV Region

Figure 7 shows that the PKED is rather flat at all times for kinetic energies lower than 0.5 eV. This clearly demonstrates that the Franck–Condon

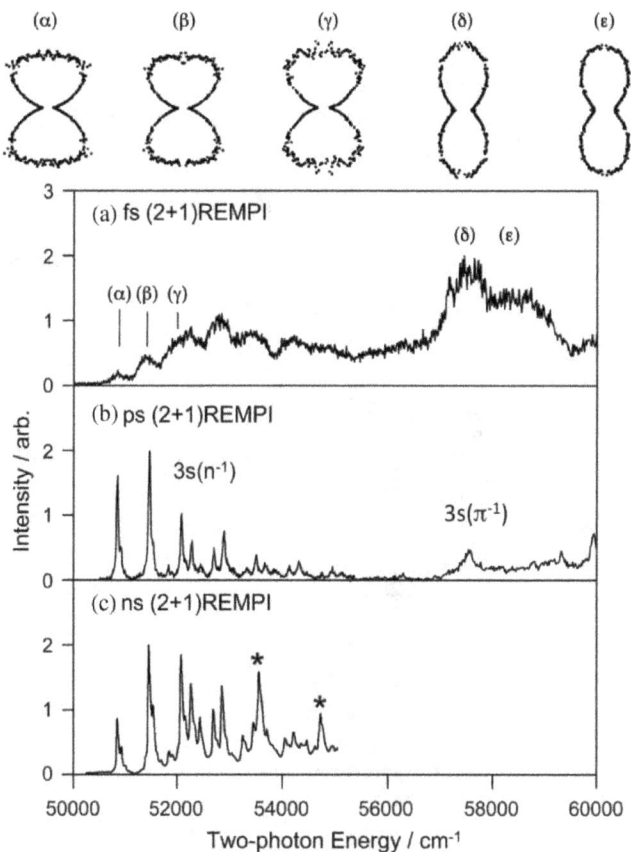

Fig. 20.   (2+1) REMPI spectra of pyrazine-h4 observed via 3s (n^{-1}) and 3s (π^{-1}) Rydberg states with (a) a femtosecond laser (150–200 fs), (b) a picosecond laser (2.8 ps), and (c) a nanosecond laser. The spectra in (a) and (c) are adapted from Refs. 67 and 68, respectively. The spectrum in (b) was measured in the present study by maintaining a constant laser power during the measurement. The spectra in (a) and (b) are of molecules in a supersonic jet, while that in (c) is of a vapor. The PADs observed for the bands (a)–(c) are shown as polar plots. The distributions are characteristic of the vibronic bands of 3s (n^{-1}) and 3s (π^{-1}) and are useful for their assignments. Reproduced with permission from Ref. 65, copyright (2008) by American Chemical Society, and Ref. 67, copyright (2001) by American Institute of Physics.

envelopes are not entirely covered for photoionization from S_2 and S_1 due to the probe photon energy being too low. VUV radiation is required to observe the entire envelopes. Femtosecond pulses in the VUV region are currently generated by at least three different methods, namely high

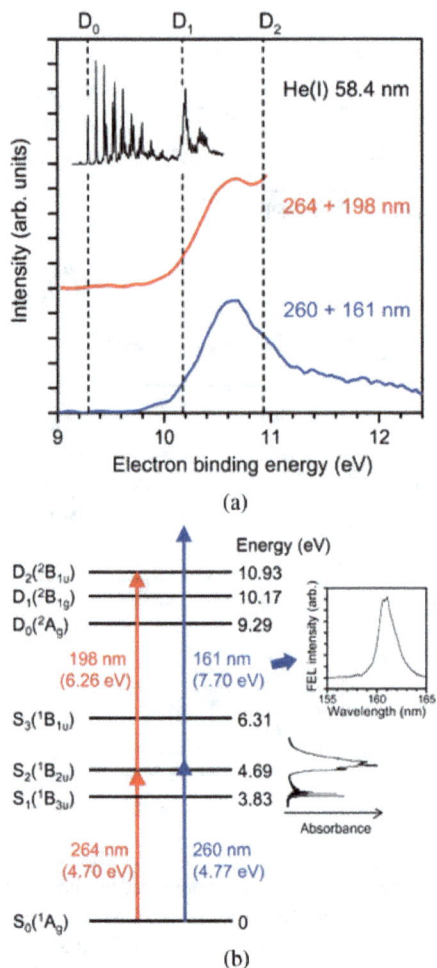

Fig. 21. (Color online) (a) PKED in He(I) photoelectron spectroscopy of ground-state pyrazine (black),[65] 264 nm pump and 198 nm probe experiment (red)[33] and 260 nm pump and 161 nm probe (blue).[79] (b) Schematic energy diagram of ionization process. Insets show UV absorption spectrum of pyrazine vapor at room temperature and time-averaged spectrum of VUV FEL. Reproduced with permission from Ref. 79, copyright (2010) by American Physical Society.

harmonic generation using an intense femtosecond laser,[69−71] free electron lasers,[72−74] and four-wave mixing.[46,75−78] We employed a VUV free electron laser (SCSS: SPring-8 Compact SASE Source)[73] to perform TRPEI experiments in combination with a femtosecond UV laser. Figure 21 compares the photoelectron spectrum measured using the 161 nm probe pulse from SCSS and the 198 nm probe pulse from filamentation deep UV source in the laboratory.[33] The influence of the pump wavelength can be neglected. The former distribution exhibits the entire Franck–Condon envelope, clearly showing a maximum in the region ca. 1.2 eV above D_0, which is consistent with the vibrational energy of ca. 0.9 eV in S_1. Since S_1 is the (n, π^*) state and D_0 and D_1 are n^{-1} and π^{-1} states, the frozen-core approximation predicts ionization occurs from S_1 to D_0, as discussed earlier. Thus, the peak of the photoelectron distribution corresponds to highly vibrationally excited levels in D_0.

Nevertheless, since SCSS uses self-amplification of spontaneous emission and a thermal cathode, its output pulse intensity, photon energy, and timing inevitably fluctuate. Consequently, the timing jitter between the SCSS and a femtosecond laser is of the order of sub-picoseconds, which does not enable us to observe $S_2 - S_1$ ultrafast internal conversion in pyrazine in real time. As explained above, filamentation four-wave mixing can generate VUV radiation by cascaded four-wave mixing. Therefore, we constructed a VUV (157 nm) filamentation light source using filamentation in Ne. The pulse energy exceeds 500 nJ at 1 kHz. All optical paths for both 264 nm and 157 nm pulses are under vacuum and only reflective optical components are used to separate and focus the two beams; consequently, output pulses from the filamentation cell are directly used for experiments without recompression to compensate the optical chirp produced as the pulses propagate through air. The cross-correlation of the 264 nm and 157 nm pulses is sub-30 fs.

Figure 22 shows the pump-probe time profile of the photoelectron intensity observed using the filamentation light source. The signal intensity is considerably higher in the negative time range where the probe pulse (157 nm) precedes the pump pulse (264 nm). This pulse order excites pyrazine to higher valence states and 3s and 3p Rydberg states and then ionizes from these states. In the positive time range, there is a flat distribution corresponding to the decay of S_1 produced by internal conversion from S_2 pumped by the 264 nm pump pulse. The high time resolution and repetition rate of this light source provides complementary experimental abilities with those of a spectroscopic system based on an intense VUV free electron laser.

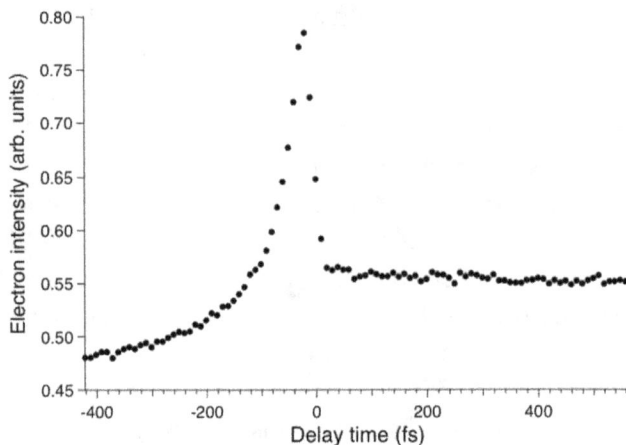

Fig. 22. Photoelectron signal time-profile observed for pyrazine using 264 nm pump and 157 nm probe pulses. Negative delay times indicate that the probe pulse precedes the pump pulse.

1.7. Summary

Ultrafast internal conversion via conical intersection plays crucial roles in the photophysics and photochemistry of aromatic molecules. In this chapter, we described how these processes can be studied in the time and frequency domains by various types of photoelectron spectroscopies. Although several theoretical studies have been performed on the $S_2 - S_1$ internal conversion in pyrazine as a benchmark system, its real-time observation was realized quite recently due to the development of sub-20 fs ultrafast lasers operating in the deep UV region. The most useful observable for studying the nonadiabatic electronic dynamics of pyrazine is the time-energy map of photoelectron anisotropy measured by photoelectron imaging. Analysis of the photoelectron kinetic energy distribution provides various data on the vibrational wave packet. Since the S_1 states populated by internal conversion from S_2 have a lifetime of only 22 ps, these states further deactivate to lower electronic states. Observation of the entire electronic deactivation process including the decay from S_1 requires pump-probe experiments using a VUV laser. Some preliminary experiments have been performed using VUV free electron lasers, while the 90 nm ultrafast lasers with 1 kHz repetition rates are currently being developed in our laboratory.

A conical intersection similar to the $S_2 - S_1$ system is also observed for the cation states of D_1 and D_0. Pulsed field ionization photoelectron spectroscopy is strongly affected by the vibronic coupling mediated by the conical intersection and its spectral feature for D_1 differs strikingly from that in He(I) photoelectron spectroscopy. Because the molecular Rydberg states have essentially the same electronic potentials as the cation states, conical intersections occur between the Rydberg states with the n^{-1} and π^{-1} ion cores. This is confirmed for the $3s(n^{-1})$ and $3s(\pi^{-1})$ Rydberg states. The complexities in the spectra of Rydberg states of pyrazine in the VUV region may be partly ascribed to conical intersections and to Rydberg-valence interactions.

Acknowledgments

The authors thank M. Oku, M. Tsubouchi, N. Nishizawa, T. Fuji, and T. Horio for their contributions to the experimental studies presented in this chapter. We express our gratitude to Professors V. Bonačić-Koutecký, R. Mitric, and C.-Y. Ng for very fruitful collaboration.

References

1. M. Born and R. Oppenheimer, *Ann. Phys.-Berlin* **84**, 0457 (1927).
2. J. von Neumann and E. Wigner, *Phys. Z* **30**, 467 (1929).
3. M. L. Hause, Y. H. Yoon and F. F. Crim, *J. Chem. Phys.* **125**, 174309 (2006).
4. F. Bernardi, M. Olivucci and M. A. Robb, *Chem. Soc. Rev.* **25**, 321 (1996).
5. W. Domcke, D. R. Yarkony, and H. Koppel, *Conical Intersections: Theory, Computation and Experiment*, Vol. 17 (World Scientific, Singapore, 2011).
6. W. Domcke, D. R. Yarkony and H. Koppel, *Conical Intersections: Electronic Structure, Dynamics & Spectroscopy*, Vol. 15 (World Scientific, Singapore, 2004).
7. D. R. Yarkony, *Acc. Chem. Res.* **31**, 511 (1998).
8. D. R. Yarkony, *Rev. Mod. Phys.* **68**, 985 (1996).
9. D. H. Mordaunt, M. N. R. Ashfold and R. N. Dixon, *J. Chem. Phys.* **104**, 6460 (1996).
10. D. H. Mordaunt, R. N. Dixon and M. N. R. Ashfold, *J. Chem. Phys.* **104**, 6472 (1996).
11. R. N. Dixon, *Mol. Phys.* **88**, 949 (1996).
12. R. N. Dixon, *Acc. Chem. Res.* **24**, 16 (1991).
13. E. Teller, *Israel J. Chem.* **7**, 227 (1969).
14. L. Salem, C. Leforestier, G. Segal and R. Wetmore, *J. Amer. Chem. Soc.* **97**, 479 (1975).
15. L. Salem, *J. Amer. Chem. Soc.* **96**, 3486 (1974).
16. R. Schneider and W. Domcke, *Chem. Phys. Lett.* **150**, 235 (1988).
17. L. Seidner, G. Stock, A. L. Sobolewski and W. Domcke, *J. Chem. Phys.* **96**, 5298 (1992).

18. C. Woywod, W. Domcke, A. L. Sobolewski and H. J. Werner, *J. Chem. Phys.* **100**, 1400 (1994).
19. T. Gerdts and U. Manthe, *Chem. Phys. Lett.* **295**, 167 (1998).
20. A. Raab, G. A. Worth, H. D. Meyer and L. S. Cederbaum, *J. Chem. Phys.* **110**, 936 (1999).
21. M. Thoss, W. H. Miller and G. Stock, *J. Chem. Phys.* **112**, 10282 (2000).
22. C. Coletti and G. D. Billing, *Chem. Phys. Lett.* **368**, 289 (2003).
23. D. V. Shalashilin and M. S. Child, *J. Chem. Phys.* **121**, 3563 (2004).
24. X. Chen and V. S. Batista, *J. Chem. Phys.* **125**, 124313 (2006).
25. P. Puzari, B. Sarkar and S. Adhikari, *J. Chem. Phys.* **125**, 194316 (2006).
26. P. Puzari, R. S. Swathi, B. Sarkar and S. Adhikari, *J. Chem. Phys.*, **123**, 134317 (2005).
27. R. X. He, C. Y. Zhu, C. H. Chin and S. H. Lin, *Chem. Phys. Lett.* **476**, 19 (2009).
28. U. Werner, R. Mitrić, T. Suzuki and V. Bonačić-Koutecký, *Chem. Phys.* **349**, 319 (2008).
29. U. Werner, R. Mitrić and V. Bonačić-Koutecký, *J. Chem. Phys.*, **132**, 174301 (2010).
30. C. K. Lin, Y. L. Niu, C. Y. Zhu, Z. G. Shuai and S. H. Lin, *Chem-Asian J.* **6**, 2977 (2011).
31. R. Lord, A. Marston and F. A. Miller, *Spectrochim. Acta* **9**, 113 (1957).
32. K. K. Innes, I. G. Ross and W. R. Moomaw, *J Mol. Spectrosc.* **132**, 492 (1988).
33. Y. I. Suzuki, T. Fuji, T. Horio and T. Suzuki, *J. Chem. Phys.* **132**, 174302 (2010).
34. I. Yamazaki, T. Murao, T. Yamanaka and K. Yoshihara, *Farad. Discuss. Chem. Soc.* **75**, 395 (1983).
35. E. J. Heller, *J. Chem. Phys.* **68**, 3891 (1978).
36. E. J. Heller, *Acc. Chem. Res.* **14**, 368 (1981).
37. M. Seel and W. Domcke, *J. Chem. Phys.* **95**, 7806 (1991).
38. S. Hahn and G. Stock, *Phys. Chem. Chem. Phys.* **3**, 2331 (2001).
39. L. Wang, H. Kohguchi and T. Suzuki, *Farad. Discuss*, **113**, 37 (1999).
40. T. Suzuki, L. Wang and H. Kohguchi, *J. Chem. Phys.* **111**, 4859 (1999).
41. V. Stert, P. Farmanara and W. Radloff, *J. Chem. Phys.* **112**, 4460 (2000).
42. T. Suzuki, Time-resolved photoelectron spectroscopy and imaging, In *Modern Trends in Chemical Reaction Dynamics: Theory and Experiment (Part I)*; C.-Y. Ng, ed. (World Scientific, Singapore, 2004).
43. I. V. Hertel and W. Radloff, *Rep. Prog. Phys.* **69**, 1897 (2006).
44. L. Misoguti, I. P. Christov, S. Backus, M. M. Murnane and H. C. Kapteyn *Phys. Rev. A* **72**, 063803 (2005).
45. L. Misoguti, S. Backus, C. G. Durfee, R. Bartels, M. M. Murnane and H. C. Kapteyn, *Phys. Rev. Lett.* **87**, 013601 (2001).
46. T. Fuji, T. Horio and T. Suzuki, *Opt. Lett.* **32**, 2481 (2007).
47. S. L. Chin, *Femtosecond Laser Filamentation* (Springer, New York, 2010).
48. E. T. J. Nibbering, G. Grillon, M. A. Franco, B. S. Prade and A. Mysyrowicz, *J. Opt. Soc. Am. B* **14**, 650 (1997).
49. D. W. Chandler and P. L. Houston, *J. Chem. Phys.* **87**, 1445 (1987).
50. A. T. J. B. Eppink and D. H. Parker, *Rev. Sci. Instrum.* **68**, 3477 (1997).
51. J. J. Lin, J. G. Zhou, W. C. Shiu and K. P. Liu, *Rev. Sci. Instrum.* **74**, 2495 (2003).
52. S. Y. Liu, K. Alnama, J. Matsumoto, K. Nishizawa, H. Kohguchi, Y. P. Lee and T. Suzuki, *J. Phys. Chem. A* **115**, 2953 (2011).
53. T. Horio and T. Suzuki, *Rev. Sci. Instrum.* **80**, 013706 (2009).

54. Y. Ogi, H. Kohguchi, D. Niu, K. Ohshimo and T. Suzuki, *J. Phys. Chem. A* **113**, 14536 (2009).
55. G. A. Garcia, L. Nahon and I. Powis, *Rev. Sci. Instrum.* **75**, 4989 (2004).
56. T. Horio, T. Fuji, Y. I. Suzuki and T. Suzuki, *J. Amer. Chem. Soc.* **131**, 10392 (2009).
57. Y. Udagawa, M. Ito and I. Suzuka, *Chem. Phys.* **46**, 237 (1980).
58. T. Fuji, Y. I. Suzuki, T. Horio and T. Suzuki, *Chem-Asian J.* **6**, 3028 (2011).
59. H. F. Schaefer, R. A. Klemm and F. E. Harris, *Phys. Rev.* **181**, 137 (1969).
60. Y. I. Suzuki and T. Suzuki, *J. Phys. Chem. A* **112**, 402 (2008).
61. M. P. Fülscher, K. Andersson and B. O. Roos, *J. Phys. Chem. A* **96**, 9204 (1992).
62. D. Holland, A. Potts, L. Karlsson, M. Stener and P. Decleva, *Chem.Phys.* **390**, 25 (2011).
63. Y. Arasaki, K. Takatsuka, K. Wang and V. Mckoy, *J. Phys. Chem. A* **132**, 124307 (2010).
64. L. Seidner, W. Domcke and W. Vonniessen, $\tilde{X}^2 A_g$-$\tilde{A}^2 B_{1g}$, *Chem. Phys. Lett.* **205**, 117 (1993).
65. M. Oku, Y. Hou, X. Xing, B. Reed, H. Xu, C. Chang, C. Y. Ng, K. Nishizawa, K. Ohshimo and T. Suzuki, *J. Phys. Chem. A* **112**, 2293 (2008).
66. C. Fridh, L. Åsbrink, B. O. Jönsson and E. Lindholm, *Int. J. Mass Spec. Ion Phys.* **8**, 101 (1972).
67. J. K. Song, M. Tsubouchi and T. Suzuki, *J. Chem. Phys.* **115**, 8810 (2001).
68. R. E. Turner, V. Vaida, C. A. Molini, J. O. Berg and D. H. Parker, *Chem. Phys.* **28**, 47 (1978).
69. T. Pfeifer, C. Spielmann and G. Gerber, *Rep. Prog. Phys.* **69**, 443 (2006).
70. F. Krausz and M. Ivanov, *Rev. Mod. Phys.* **81**, 163 (2009).
71. M. Nisoli and G. Sansone, *Prog. Quant. Electron.* **33**, 17 (2009).
72. V. Ayvazyan, N. Baboi, J. Bahr, V. Balandin, B. Beutner, A. Brandt, I. Bohnet, A. Bolzmann, R. Brinkmann, O. I. Brovko, J. P. Carneiro, S. Casalbuoni, M. Castellano, P. Castro, L. Catani, E. Chiadroni, S. Choroba, A. Cianchi, H. Delsim-Hashemi, G. Di Pirro, M. Dohlus, S. Dusterer, H. T. Edwards, B. Faatz, A. A. Fateev, J. Feldhaus, K. Flöttmann, J. Frisch, L. Fröhlich, T. Garvey, U. Gensch, N. Golubeva, H. J. Grabosch, B. Grigoryan, O. Grimm, U. Hahn, J. H. Han, M. V. Hartrott, K. Honkavaara, M. Hüning, R. Ischebeck, E. Jaeschke, M. Jablonka, R. Kammering, V. Katalev, B. Keitel, S. Khodyachykh, Y. Kim, V. Kocharyan, M. Körfer, M. Kollewe, D. Kostin, D. Krämer, M. Krassilnikov, G. Kube, L. Lilje, T. Limberg, D. Lipka, F. Löhl, M. Luong, C. Magne, J. Menzel, P. Michelato, V. Miltchev, M. Minty, W. D. Möller, M. Monaco, W. Müller, M. Nagl, O. Napoly, P. Nicolosi, D. Nölle, T. Nuñez, A. Oppelt, C. Pagani, R. Paparella, B. Petersen, B. Petrosyan, J. Pflüger, P. Piot, E. Plönjes, L. Poletto, D. Proch, D. Pugachov, K. Rehlich, D. Richter, S. Riemann, M. Ross, J. Rossbach, M. Sachwitz, E. L. Saldin, W. Sandner, H. Schlarb, B. Schmidt, M. Schmitz, P. Schmuser, J. R. Schneider, E. A. Schneidmiller, H. J. Schreiber, S. Schreiber, *Eur. Phys. J. D* **37**, 297 (2006).
73. T. Shintake, H. Tanaka, T. Hara, T. Tanaka, K. Togawa, M. Yabashi, Y. Otake, Y. Asano, T. Bizen, T. Fukui, S. Goto, A. Higashiya, T. Hirono, N. Hosoda, T. Inagaki, S. Inoue, M. Ishii, Y. Kim, H. Kimura, M. Kitamura, T. Kobayashi, H. Maesaka, T. Masuda, S. Matsui, T. Matsushita, X. Maréchal, M. Nagasono, H. Ohashi, T. Ohata, T. Ohshima, K. Onoe, K. Shirasawa, T. Takagi, S. Takahashi, M. Takeuchi, K. Tamasaku, R. Tanaka, Y. Tanaka, T. Tanikawa, T. Togashi, S. Wu, A. Yamashita, K. Yanagida, C. Zhang, H. Kitamura and Ishikawa, T, *Nat. Photon.* **2**, 555 (2008).

74. A. Barty, R. Soufli, T. McCarville, S. L. Baker, M. J. Pivovaroff, P. Stefan and R. Bionta, *Opt. Exp.* **17**, 15508 (2009).

75. P. Zuo, T. Fuji, T. Horio, S. Adachi and T. Suzuki, *Appl. Phys. B*, **108**, 815 (2012).

76. M. Ghotbi, M. Beutler and F. Noack, *Opt. Lett.* **35**, 3492 (2010).

77. M. Beutler, M. Ghotbi, F. Noack and I. V. Hertel, *Opt. Lett.* **35**, 1491 (2010).

78. M. Beutler, M. Ghotbi and F. Noack, *Opt. Lett.* **36**, 3726 (2011).

79. S. Y. Liu, Y. Ogi, T. Fuji, K. Nishizawa, T. Horio, T. Mizuno, H. Kohguchi, M. Nagasono, T. Togashi, K. Tono, M. Yabashi, Y. Senba, H. Ohashi, H. Kimura, T. Ishikawa and T. Suzuki, *Phys. Rev. A* **81**, 031403 (2010).

CHAPTER 5

QUANTUM DYNAMICS IN DISSIPATIVE MOLECULAR SYSTEMS

Hou-Dao Zhang*, J. Xu*, Rui-Xue Xu† and Y. J. Yan*,†

1. Introduction

Quantum dissipation theory governs the dynamics of a quantum system embedded in a quantum bath environment. The latter has an enormous number of degrees of freedom and is subject to a quantum statistical mechanics treatment. The system of primary interest is often of a small dimension and described by the reduced system density operator, $\rho \equiv \text{tr}_B \rho_{\text{total}}$, i.e. the partial trace of the total system-plus-bath density operator over the bath subspace. The influence of bath causes many phenomena, including not just the system energy relaxation and decoherence processes, but also the particle and even the quantum information exchanges between system and environment.

Dissipation is inevitable in condensed phase systems and plays a crucial role in many fields of science. Particular interests are those of

*Department of Chemistry, Hong Kong University of Science and Technology, Hong Kong.
†Hefei National Laboratory for Physical Sciences at the Microscale, University of Science and Technology of China, China.

structured bath environment, such as proteins in photosynthesis antenna complexes, where excitation energy transfer is observed to exhibit long-lived quantum coherence even at room temperature.[1-5] Photosynthesis complexes are nano-structured systems, in which the coupling between pigment and protein has the same magnitude as that between pigments and the timescale of the protein environment memory is comparable to that of the energy transfer.[6,7] The aforementioned structural-dynamical characteristics are rather common in nano-systems and often play crucial roles in determining the underlying mechanisms of process and functionality. These characteristics resemble the *Goldilocks Principle* for a "just-right" level of complexity to optimize both function and robustness. In quantum regime this complexity principle should also include a collaboration between coherent evolution and environmental fluctuations. Apparently, traditional perturbative and Markovian quantum dissipation theories[8] are largely inadequate for these complex systems. Nonperturbative and non-Markovian approaches are in need. Among them, the hierarchical equations of motion (HEOM) approach[9-14] has emerged as a standard method. In particular, it has been used extensively in the study of light-harvesting photosynthesis systems.[15-19]

This chapter reviews our recent advancement on the HEOM formalism. The related background on the HEOM construction will be presented in Sec. 2, followed by remarks in Sec. 3 on the key features and challenges of the formalism. In Sec. 4, we discuss the optimized HEOM theory with accuracy control, constructed based on the so-called Padé spectrum decomposition (PSD) scheme[20,21] that addresses some key issues on the aforementioned minimum statistical bath basis set. In Sec. 5, we validate HEOM as a fundamental theory in quantum mechanics for open systems. We will see that the HEOM formalism defines a linear space, the HEOM space, which naturally supports the Schrödinger picture, Heisenberg picture, and interaction picture for dissipative dynamics of open systems. In conjunction of efficient evaluation of nonlinear optical response functions, we further discuss the mixed Heisenberg–Schrödinger picture, combining with a block-matrix implementation of the underlying HEOM dynamics. Applications of HEOM are exemplified in Sec. 6 with the simulation of coherent two-dimensional spectroscopy signals of model trimer systems. Concluding remarks are given in Sec. 7.

2. HEOM versus Path Integral Formalism: Background

2.1. *Generic form and terminology of HEOM*

To have a brief introduction of the HEOM formalism and terminology, let us consider its generic form[13]:

$$\dot{\rho}_n(t) = -[i\mathcal{L}(t) + \gamma_n + \delta\mathcal{R}_n]\rho_n(t) + \rho_n^{\{+\}}(t) + \rho_n^{\{-\}}(t). \qquad (2.1)$$

The reduced system Liouvillian $\mathcal{L}(t)$ can be time-dependent, e.g., in the case of pulsed laser interaction. The dynamics quantities of the formalism are a set of so-called auxiliary density operators (ADOs), each of them has the labeling index of $n \equiv \{n_1, \ldots, n_K\}$, so that $\rho_n \equiv \rho_{n_1,\ldots,n_K}$, with $\{n_k \geq 0; k = 1, \ldots, K\}$, in the case of bosonic bath interaction. We call ρ_n an nth-tier ADO, assuming its labeling index satisfies $n_1 + \cdots + n_K = n$. The reduced system density operator is just the zeroth-tier ADO, i.e., $\rho(t) = \rho_0(t)$. The last two terms in Eq. (2.1) describe how a specified nth-tier ρ_n depends on its associated $(n \pm 1)$th-tier ADOs. We will see the fact that the ADO's index n consists of K subindexes arises from decomposing the interaction bath correlation functions into distinct K exponential components. This will be referred as the bath memory-frequency decomposition, as the involving exponents can be complex in general. The explicit HEOM expressions are also dictated by this decomposition. In other words, the bath memory-frequency decomposition scheme serves as the *statistical bath basis set* for an explicit HEOM construction. The basis set size K or the number of distinct memory-frequency components amounts to the dimension of the ADO's index n. The complex damping parameter γ_n in Eq. (2.1) assumes the form of $\gamma_n = \sum_k n_k \gamma_k$ that collects all relevant complex exponents to the specified ADO. Included in the formalism is also a residue dissipation superoperator, $\delta\mathcal{R}_n$, for a partial resum of the residue dissipation outside the finite basis set of size K, in relation to the HEOM construction.[13]

2.2. *Statistical mechanics description of bath influence*

We shall be interested in the reduced dynamics of a system embedded in a bosonic bath environment. The total system-and-bath composite Hamiltonian assumes $H(t) + h_B - \sum_a \hat{Q}_a \hat{F}_a$, which in the bath h_B-interaction picture reads

$$H_{\text{total}}(t) = H(t) - \sum_a \hat{Q}_a \hat{F}_a(t). \tag{2.2}$$

The reduced system Hamiltonian $H(t)$, which enters the HEOM formalism (Eq. (2.1)) via the Liouvillian, $\mathcal{L}(t) \cdot \equiv [H(t), \cdot]$, is rather general and can contain arbitrary anharmonicity and time-dependent external field action. Throughout the paper, we set $\hbar = 1$ and $\beta \equiv 1/(k_B T)$, with the Boltzmann constant k_B and temperature T.

The last term in Eq. (2.2) denotes the system-bath coupling in a stochastic description. The involving system operator $\hat{Q}_a$ is rather general. It is called a dissipative mode, through which a generalized Langevin force $\hat{F}_a(t) \equiv e^{ih_B t} \hat{F}_a e^{-ih_B t}$ acts on the system. Apparently, the system dissipative mode $\hat{Q}_a$ characterizes the nature of dissipation, either energy relaxation or decoherence or both. In particular, a pure dephasing mode $\hat{Q}_a$ commutes with the reduced system Hamilton ion, thus does not cause the system energy change. Without loss of generality, we set the dissipative mode $\hat{Q}_a$ be dimensionless, while the generalized Langevin force is of $\langle F_a \rangle_{\text{B}} \equiv \text{tr}_{\text{B}}(F_a \rho_{\text{B}}^{\text{eq}}) = 0$, where $\rho_{\text{B}}^{\text{eq}} = e^{-\beta h_B}/\text{tr}_{\text{B}} e^{-\beta h_B}$ is the thermal equilibrium density operator of the bare bath. Assume further the stochastic bath operators $\{\hat{F}_a(t)\}$ obey the Gaussian statistics. The influence of bath, according to the Wick's theorem for thermodynamics average, can be completely characterized by the bath correlation functions[8,22]:

$$C_{ab}(t - \tau) \equiv \langle \hat{F}_a(t) \hat{F}_b(\tau) \rangle_{\text{B}}. \tag{2.3}$$

It satisfies

$$C_{ab}^*(t) = C_{ba}(-t) = C_{ab}(t - i\beta). \tag{2.4}$$

The related spectral density function is defined as

$$J_{ab}(\omega) \equiv \frac{1}{2} \int_{-\infty}^{\infty} dt\, e^{i\omega t} \langle [\hat{F}_a(t), \hat{F}_b(0)] \rangle_{\text{B}}. \tag{2.5}$$

It satisfies $J_{ab}(\omega) = -J_{ba}(-\omega) = J_{ba}^*(\omega)$, in line with the first identity of Eq. (2.4). Note that Eq. (2.5) generalizes the Caldeira–Leggett's system-bath coupling model,[23] in which $\hat{F}_a = \sum_j c_{aj} x_j$ and $h_B = \frac{1}{2} \sum_j \omega_j (x_j^2 + p_j^2)$, resulting in the spectral density of $J_{ab}(\omega) = \frac{1}{2}\pi \sum_j c_{aj} c_{bj} [\delta(\omega - \omega_j) -$

$\delta(\omega + \omega_j)]$. Note also that the second identity of Eq. (2.4) is the detailed-balance relation in the time-domain. It together with Eq. (2.5) leads to

$$C_{ab}(t) = \frac{1}{\pi} \int_{-\infty}^{\infty} d\omega \frac{e^{-i\omega t} J_{ab}(\omega)}{1 - e^{-\beta\omega}}. \tag{2.6}$$

This is the fluctuation–dissipation theorem for bosonic canonical ensembles. It will be exploited later in the exponential expansion of bath correlation function $C_{ab}(t)$, as required by the HEOM construction via the Feynman–Vernon influence functional path integral theory.

2.3. *Feynman–Vernon influence functional formalism*

Let $\mathcal{U}(t, t_0)$ be the reduced Liouville-space propagator, by which the reduced density operator $\rho(t) \equiv \text{tr}_B \rho_{\text{total}}(t)$ at time t is related to its initial value at time t_0 via

$$\rho(t) \equiv \mathcal{U}(t, t_0)\rho(t_0). \tag{2.7}$$

In a path integral formalism of quantum dynamics of open systems, the subspace of reduced system should be assigned with a specific representation. We denote $\{|\psi\rangle\}$ as a generic basis set, and $\boldsymbol{\psi} \equiv (\psi, \psi')$. Therefore, $\rho(\boldsymbol{\psi}, t) \equiv \rho(\psi, \psi', t) \equiv \langle\psi|\rho(t)|\psi'\rangle$. Denote also $Q_a[\psi(t)]$ and $Q_a[\psi'(t)]$ for the path integral representations of system dissipative mode $\hat{Q}_\alpha$ acting on the ket and bra sides (or the forward and backward paths), respectively. The reduced Liouville–space propagator in the path integral formulation can be expressed as[24]

$$\mathcal{U}(\boldsymbol{\psi}, t; \boldsymbol{\psi}_0, t_0) = \int_{\psi_0[t_0]}^{\psi[t]} \mathcal{D}\boldsymbol{\psi} \, e^{iS[\psi]} \mathcal{F}[\boldsymbol{\psi}] \, e^{-iS[\psi']}. \tag{2.8}$$

Here, $S[\psi]$ is the classical action functional of the reduced system, evaluated along a path $\psi(\tau)$, subject to the constraint that the two ending points $\psi(t_0) = \psi_0$ and $\psi(t) = \psi$ are fixed. As a result, in the absence of dissipation or $\mathcal{F}[\boldsymbol{\psi}] = 1$, the time derivative of Eq. (2.8) results in $\partial_t \mathcal{U} = -i[H(t), \mathcal{U}] \equiv -i\mathcal{L}\mathcal{U}$, which is equivalently the von Neumann equation $\dot{\rho} = -i\mathcal{L}\rho$ at the operator level. The key quantity in Eq. (2.8) is the Feynman–Vernon influence functional $\mathcal{F}[\boldsymbol{\psi}]$. In contact with the HEOM formalism to be presented later, it is written in terms of the dissipation functional, $\mathcal{R}[\tau; \{\boldsymbol{\psi}\}]$, as[12,13,25]

$$\mathcal{F}[\boldsymbol{\psi}] = \exp\left\{ -\int_{t_0}^{t} d\tau\, \mathcal{R}[\tau; \{\boldsymbol{\psi}\}] \right\}. \qquad (2.9)$$

Here

$$\mathcal{R}[t; \{\boldsymbol{\psi}\}] = i \sum_{a} \mathcal{A}_a[\boldsymbol{\psi}(t)]\,\mathcal{B}_a[t; \{\boldsymbol{\psi}\}], \qquad (2.10)$$

with

$$\mathcal{A}_a[\boldsymbol{\psi}(t)] = Q_a[\psi(t)] - Q_a[\psi'(t)], \qquad (2.11)$$

$$\mathcal{B}_a[t; \{\boldsymbol{\psi}\}] = -i\{B_a[t; \{\psi\}] - B_a'[t; \{\psi'\}]\} \qquad (2.12)$$

and

$$\mathcal{B}_a[t; \{\psi\}] \equiv \sum_{b} \int_{t_0}^{t} d\tau\, C_{ab}(t - \tau)\, Q_b[\psi(\tau)],$$

$$\qquad (2.13)$$

$$\mathcal{B}_a'[t; \{\psi'\}] \equiv \sum_{b} \int_{t_0}^{t} d\tau\, C_{ab}^{*}(t - \tau)\, Q_b[\psi'(\tau)].$$

Here, $C_{ab}(t - \tau)$ denotes the bath correlation function [Eq. (2.3)] that satisfies the fluctuation–dissipation theorem of Eq. (2.6) for bosonic canonical ensembles.

Direct implementation of the path integral formulations is very expensive, even with the forward–backward iterative propagation method[26,27] or its advanced variations. We will show that the equivalent HEOM theory, Eq. (2.1), to be constructed in Sec. 4, on the basis of exponential expansion of bath correlation function (see Sec. 3), is not just much more efficient numerically, but also operational friendly (see Sec. 5).

2.4. *General comments*

To conclude this section, let us discuss some general features that are of fundamental significance for the path integral formalism and its equivalent HEOM theory.

(1) *About the initial system-bath correlation.* In the derivation of above influence functional path integral formalism, we adopt the initial factorization ansatz of $\rho_T(t_0) = \rho(t_0)\rho_B^{eq}$. This ansatz can in principle be exact, provided that the initial time, t_0, corresponds to infinitely remote past, i.e., $t_0 \to -\infty$.[8,22,28] In practice, the initial time for a physical dynamic process is often set as $t_0 = 0$ from the moment right after external field acts. It is important to distinguish $t_0 \to -\infty$ from $t_0 = 0$, as the former is taken as a reference time instant, while the latter is associated with a physical state where the system and bath is fully coupled. The system-bath correlation at $t_0 = 0$ is important to the subsequent reduced system dynamics. This initial system-bath correlation at thermal equilibrium is accounted for by the nonzero ADOs in the stationary solution, i.e., $\{\dot{\rho}_n^{eq} = 0\}$, to HEOM, Eq. (2.1).

(2) *Time local versus memory functionals.* The dissipation functional $\mathcal{R}[\tau; \{\psi\}]$ in Eq. (2.10) is decomposed, similar to the decomposition of system-bath coupling in the last term of Eq. (2.2), as the sum of contributions from individual dissipative modes. Each term consists of the product of composite $\mathcal{A}[\psi(t)]$- and $\mathcal{B}[t; \{\psi\}]$-type functionals, which are different in their memory contents and time orderings. The $\mathcal{A}$-type functionals [Eq. (2.11)] are just the individual dissipative modes in the path integral representation, and depend only on the fixed ending point $\psi(t) = (\psi, \psi')$ of the path. Consequently, the operator level expression of $\mathcal{A}_a[\psi(t)]$ by Eq. (2.11) is straightforward. It maps to the commutator as $\mathcal{A}_a\hat{O} = [\hat{Q}_a, \hat{O}]$, for its action on an arbitrary operator $\hat{O}$. In contrast, the $\mathcal{B}$-type functionals [Eq. (2.12)] depend on the bath correlation functions and thus contain memory on all previous paths $\psi(\tau)$, with $t_0 \le \tau \le t$. Therefore, they do not have explicit correspondence at the operator level. Moreover, as the time ordering is concerned, the $\mathcal{B}$-type functionals proceed prior to the $\mathcal{A}$-type functionals in path integral.

(3) *Physical time ordering.* Consider now a generic product term $\mathcal{A}_a\mathcal{B}_a$, in Eq. (2.10), where both $\mathcal{A}_a[\psi(t)]$ (Eq. (2.11)) and $\mathcal{B}_a[t; \{\psi\}]$ (Eq. (2.12)) are c-numbers. Their product has however the physically preferred ordering. Note that $\mathcal{A}_a[\psi(t)] = Q_a[\psi(t)] - Q_a[\psi'(t)]$ (cf. Eq. (2.11)), where $\psi(\tau)$ and $\psi'(\tau)$ denote the forward and backward paths in the integral evolution, for the left ket and right bra sides of ρ, respectively. Thus, as the time

ordering is considered, the product $\mathcal{A}_a \mathcal{B}_a$ of c-numbers reads

$$\mathcal{A}_a[\psi(t)]\mathcal{B}_a[t; \{\psi\}] = \mathcal{Q}_a[\psi(t)]\mathcal{B}_a[t; \{\psi\}] - \mathcal{B}_a[t; \{\psi\}]\mathcal{Q}_a[\psi'(t)].$$

$$(2.14)$$

Note that the time derivative on influence functional is given by

$$\partial_t \mathcal{F}[\psi] = -\mathcal{R}[t; \{\psi\}]\mathcal{F}[\psi] = -i \sum_a \mathcal{A}_a[\psi(t)]\mathcal{B}_a[t; \{\psi\}]\mathcal{F}[\psi],$$

$$(2.15)$$

as inferred from Eqs. (2.9) and (2.10).

3. Memory-Frequency Decomposition of Bath Correlation Functions

3.1. *PSD of Bose function*

The Feynman–Vernon path integral formalism (Eqs. (2.7)–(2.13)) is formally exact, provided the generalized Langevin forces $\{\hat{F}_a(t)\}$ from bath satisfy the Gaussian statistics. The resulting integral equation for the dissipative propagator $\mathcal{U}(t, t_0)$ is however difficult for numerical treatment. To transform it into a set of linearly coupled differential equations, HEOM, which are much more convenient for numerical calculation, we expand the bath correlation functions $C_{ab}(t)$ in exponential series form (see Eq. (4.2)), so that the time derivative on each exponential term leads to itself. The time derivatives on the influence functional and its hierarchical auxiliaries will then lead readily to the HEOM formulations for nonperturbative quantum dissipation theory; see Sec. 4.

Note that the dissipation functional, $\mathcal{R}[\tau; \{\psi\}]$ of Eq. (2.10), is additive with respect to different components of bath correlation functions. Without loss of generality, we can hereafter focus on the single dissipative mode case, so that the index a can be omitted for simplicity. In other words, we consider the system-bath coupling the form of $H_{sb}(t) = -\hat{Q}\hat{F}(t)$, with the bath correlation function, $C(t) = \langle \hat{F}(t)\hat{F}(0)\rangle_B$, being expressed as

$$C(t) = \frac{1}{\pi} \int_{-\infty}^{\infty} d\omega \frac{e^{-i\omega t} J(\omega)}{1 - e^{-\beta\omega}} = \sum_{k=1}^{K} c_k e^{-\gamma_k t} + \delta C_N(t). \qquad (3.1)$$

The exponential series term is obtained via the Cauchy residue theorem of contour integration in the lower-half plane, based on certain sum-over-poles schemes, with a total of $K = N + N'$ poles, N from Bose function, $f^{\text{Bose}}(\omega) = \frac{1}{1-e^{-\beta\omega}}$, and N' from bath spectral density function, $J(\omega)$. Assume also that the latter has no approximation; therefore, the residue bath correlation function, $\delta C_N(t)$, arises only from the finite sum-over-poles approximation to Bose function. Adopted conventionally are the Matsubara expansion on Bose function, while the Meier–Tannor parametrization on the bath spectral density function.[8,29] Advanced sum-over-poles schemes for both functions have been developed in conjunction with optimized HEOM theory.[20,21,30,31]

Consider first sum-over-poles schemes for Bose function (set $x \equiv \beta\omega$),

$$\frac{1}{1 - e^{-x}} = \frac{1}{2} + \frac{1}{2}\frac{\cosh(x/2)}{\sinh(x/2)}. \tag{3.2}$$

The conventional approach is the Matsubara spectrum decomposition (MSD), i.e., the Matsubara expansion with finite terms,

$$\frac{1}{1 - e^{-x}} \approx f_N^{\text{MSD}}(x) = \frac{1}{2} + \frac{1}{x} + \sum_{m=1}^{N} \frac{2x}{x^2 + (2\pi m)^2}. \tag{3.3}$$

The involving Matsubara frequencies, $\{2\pi m/\beta; \; m = 1, 2, \ldots, N\}$, arise from the poles of the denominator function of $\sinh(\beta\omega/2)$ in Eq. (3.2). While it is exact if $N \to \infty$, MSD is notorious of slow convergence. To address this issue, let us recast Bose function as

$$\frac{1}{1 - e^{-x}} \equiv \frac{1}{2} + \frac{1}{x} + x\Phi(x^2) \tag{3.4}$$

and focus on the function $\Phi(y)$, with $y \equiv x^2$. Apparently, the MSD of Eq. (3.3) amounts mathematically to a polynomial fractional expression of

$$\Phi_N^{\text{MSD}}(y) = \frac{P_M(y)}{Q_N(y)} = \frac{p_0 + p_1 y + \cdots + p_M y^M}{q_0 + q_1 y + \cdots + q_N y^N}, \tag{3.5}$$

with $M = N - 1$. For an optimal sum-over-poles approximant of Bose function, we exploit the fact that the *best* approximation of a function by the specified order of polynomial fractional expression is the $[M/N]$ Padé approximant.[32] The involving $(M + N + 1)$ independent parameters are uniquely determined by the $(M + N + 1)$ coefficients of the Taylor expansion, $\Phi(y) \approx \sum_{k=0}^{M+N} a_k y^k$. Thus, it is accurate up to the order of $\mathcal{O}(y^{M+N})$. It also partially accounts for the higher-order contributions; thus mostly a Padé approximant behaves better than its Taylor series, and may even work where the Taylor series of a function does not converge. To obtain the sum-over-poles form, the roots of its Padé denominator polynomial $Q_N(y)$ should be determined with high precision. The direct method is applicable only for small N, but numerically challenging and inaccurate for high-order polynomials in general. To overcome this problem, the PSD algorithm for both Bose and Fermi functions is developed.[20,21]

In the following, we highlight the results of PSD approximants for Bose function. Let $f^{[M/N]}(x)$ be the resulting Padé approximant of Bose function. For a general M, the poles of $f^{[M/N]}(x)$ are complex. However, we have shown that the following three having only pure imaginary poles are classified as the PSD schemes[21]:

$$f^{[N-1/N]}(x) = \frac{1}{2} + \frac{1}{x} + \sum_{m=1}^{N} \frac{2\tilde{\eta}_m x}{x^2 + \tilde{\xi}_m^2}, \tag{3.6}$$

$$f^{[N/N]}(x) = \frac{1}{2} + \frac{1}{x} + \sum_{m=1}^{N} \frac{2\eta_m x}{x^2 + \xi_m^2} + R_N x, \tag{3.7}$$

$$f^{[N+1/N]}(x) = \frac{1}{2} + \frac{1}{x} + \sum_{m=1}^{N} \frac{2\check{\eta}_m x}{x^2 + \check{\xi}_m^2} + \check{R}_N x + \check{T}_N x^3. \tag{3.8}$$

They are exact up to $\mathcal{O}(x^{4N-1})$, $\mathcal{O}(x^{4N+1})$, and $\mathcal{O}(x^{4N+3})$, respectively. All the involving PSD parameters in Eqs. (3.6)–(3.8) can be evaluated readily with high precision.[21] In particular, the PSD pole ξ-parameters are all real and positive, and can therefore be used to define the PSD frequencies, in analogue with the Matsubara frequencies. With a PSD approximant, Bose function contributes to the bath correlation function $C(t)$ of Eq. (3.1) N exponential terms. The exponents are just the PSD frequencies, such as $\{\gamma_m = \xi_m/\beta; m = 1, \ldots, N\}$, in relation to the pole parameters. Note that a bosonic spectral density is an odd function,[8] i.e., $J(-\omega) = -J(\omega)$. Therefore, the coefficients $\{c_m = -2i\eta_m J(-i\gamma_m)/\beta\}$ associating with PSD frequencies in Eq. (3.1) are all real.

Note that MSD of Eq. (3.3) has similar expression as $[N-1/N]$–PSD of Eq. (3.6), but with $\eta_m^{\text{MSD}} = 1$, and $\xi_m^{\text{MSD}} = 2\pi m$. The MSD coefficients and poles are independent of N. In contrast, those of PSDs are of N-dependence. The PSD can be considered as a corrected finite truncation of MSD, tailoring primarily those poles and coefficients near the terminal from the MSD counterparts. We have also shown[20,21] that the accuracy length of each individual PSD scales as N^2, while that of MSD is only of a weak N-dependence that is responsible for its notoriously slow convergence. The superiority of PSD over the conventional MSD is shown to be remarkably significant.

Both $[N/N]$ and $[N+1/N]$ of Bose function diverge at large x, while $[N-1/N]$ approaches to x^{-1}. The divergence parameter R_N in $[N/N]$ has the expression $R_N = [4(N+1)(2N+3)]^{-1}$, but those for $\check{R}_N$ and $\check{T}_N$ in $[N+1/N]$ are rather lengthy.[21] We find that $\check{R}_N$ and R_N behave similarly, while $\check{T}_N$ is negative and scales approximately as $\check{T}_N \propto N^{-6}$, for large N.[31] We may make use of the divergence of the given Bose function approximant to achieve the white-noise residue limit, i.e., $\delta C_N(t) \propto \delta(t)$ in Eq. (3.1). It would lead to an optimized HEOM, as elaborated in Sec. 4.

3.2. *Brownian oscillators decomposition of bath spectral density function*

Turn now to the sum-over-poles schemes on bath spectral density $J(\omega)$. It is noticed that the Meier–Tannor parametrization method[8,29] was often adopted in the development of perturbative quantum dissipation theory. This scheme covers the range of underdamped (including the critically damped) bath modes, but not the case of overdamped bath. At any rate, perturbative theories are completely inapplicable in the former case. However, the HEOM theory is exact for arbitrary bath interactions, as long as they satisfy the Gaussian statistics. It is equivalent to Feynman–Vernon influence functional formalism. In fact the HEOM dynamics with the inclusion of underdamped bath modes can reveal clearly the correlated system-bath coherence.[33]

For completeness and also to be in contact with molecular reality, we present below the multiple Brownian oscillators (BOs) model for the sum-over-poles parametrization on bath spectral density function. It reads

$J(\omega) = \sum_\alpha J_\alpha(\omega)$, with each $J_\alpha(\omega)$ the BO's form of Refs. 8, 22 and 31

$$J(\omega) = \frac{2\lambda\omega_{BO}^2 \zeta_{BO}\omega}{(\omega_{BO}^2 - \omega^2)^2 + \zeta_{BO}^2\omega^2}. \qquad (3.9)$$

Here $\lambda = \int d\omega J(\omega)/(2\pi\omega)$ is contribution of each bath mode to overall reorganization energy that is proportional to the system-bath coupling strength.[8,22] The Huang–Rhys factor amounts to λ/ω_{BO} in the damping-free oscillator limit. Each BO has two poles in the lower-half plane, and thus contributes, in general, two exponential terms to $C(t)$ in Eq. (3.1), with the exponents of $\gamma_{BO}^\pm \in \{\gamma_k\}$ being

$$\gamma_{BO}^\pm = \frac{1}{2}\zeta_{BO} \pm i\left(\omega_{BO}^2 - \frac{1}{4}\zeta_{BO}^2\right)^{1/2}. \qquad (3.10)$$

The corresponding coefficients of $c_{BO}^\pm \in \{c_k\}$ would read then $c_{BO}^\pm = -2i[(z+i\gamma_{BO}^\pm)f^{Bose}(z)J(z)]_{z=-i\gamma_{BO}^\pm}$, where $f^{Bose}(z)$ stands for Bose function, $\frac{1}{1-e^{-\beta z}}$, appearing in Eq. (3.1). However, the expansion of bath correlation function in the form of Eq. (3.1) will also involve certain sum-of-poles approximant for Bose function; see Sec. 4.1. Thus, the exponential expansion coefficients are usually better to be evaluated with an approximated value rather than the exact Bose function in Eq. (3.1).

An individual BO bath mode can be characterized by

$$r_{BO} \equiv \left(\frac{1}{2}\zeta_{BO}/\omega_{BO}\right)^2. \qquad (3.11)$$

It specifies the strongly underdamped ($r_{BO} < 0.5$), weakly underdamped ($0.5 \le r_{BO} < 1$), critically damped ($r_{BO} = 1$), overdamped ($r_{BO} > 1$) BO cases, respectively. Included here is also the special value of $r_{BO} = 0.5$, at which Re $\gamma_{BO}^\pm$ and Im $\gamma_{BO}^\pm$ are of same amplitude, see Eq. (3.10). It features rather the property of $J(\omega)/\omega$, i.e., the spectrum of frictional kernel,[8] which is an even function and plotted in Fig. 1. The friction spectrum $J(\omega)/\omega$ has two symmetric peaks when $r_{BO} < 0.5$, which are merged into one at $\omega = 0$ when $r_{BO} \ge 0.5$. This feature is closely related to the construction of an optimized HEOM theory.[31] In particular, we have proved that for the case of all bath modes being of $r_{BO} \ge 0.5$, the optimized HEOM construction goes with the $[N+1/N]$-PSD approximant of Bose function.

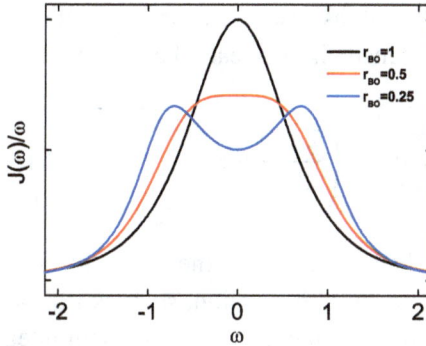

Fig. 1. (Color online) Friction spectrum $J(\omega)/\omega$, via Eq. (3.9) by a common reorganization energy λ, as function of ω (in unit of ω_{BO}), for the critically damped ($r_{BO} = 1$, black), critically-weakly underdamped ($r_{BO} = 0.5$, red), and strongly underdamped ($r_{BO} < 0.5$, blue) BO cases, respectively.

4. Optimized HEOM Theory With Accuracy Control

4.1. *Construction of HEOM via path integral formalism*

In this subsection, we exemplify the HEOM construction with the Drude bath model of

$$J(\omega) = \frac{2\lambda\gamma_D\omega}{\omega^2 + \gamma_D^2}. \tag{4.1}$$

This is the strongly overdamped or Smoluchowski limit of Eq. (3.9), with $r_{BO} \gg 1$, or more precisely, $\zeta_{BO} \gg (\omega_{BO}; \omega)$ for the entire frequency range of interest, but $\gamma_D \equiv \omega_{BO}^2/\zeta_{BO}$ being finite. This model has only one pole ($z = -i\gamma_D$) in the lower-half plane. Combining with the $[N/N]$–PSD approximant for Bose function in Eq. (3.1), we obtain the Drude bath correlation function of the form

$$C(t) \approx \sum_{k=1}^{K} c_k e^{-\gamma_k t} + 2\Delta_N\delta(t). \tag{4.2}$$

The associated white-noise residue is of $\Delta_N = 2\lambda\beta\gamma_D R_N = \frac{\lambda\beta\gamma_D}{2(N+1)(2N+3)}$. In Eq. (4.2), $K = N + 1$. The first N exponents and coefficients, $\{\gamma_k$ and $c_k; k = 1, \ldots, N\}$, arising from the poles of $[N/N]$ Bose function approximant are all real, as discussed earlier, while $\gamma_{N+1} = \gamma_{\mathrm{D}}$ from the Drude bath spectral density is also real. The Drude coefficient is found to be

$$c_{N+1} = c_{\mathrm{D}} = \left(\frac{2\lambda}{\beta} - \gamma_{\mathrm{D}} \Delta_N - \gamma_{\mathrm{D}} \sum_{j=1}^{N} \frac{c_j}{\gamma_j} \right) - i\lambda\gamma_{\mathrm{D}}. \tag{4.3}$$

We will show below that with the bath correlation function being expressed in an exponential expansion, such as Eq. (4.2), HEOM can be derived readily from the influence functional path integral theory. For the single mode case in consideration, the dissipation functional of Eq. (2.10) reads

$$\mathcal{R}[\tau; \{\boldsymbol{\psi}\}] = i\mathcal{A}[\boldsymbol{\psi}(t)]\mathcal{B}[t; \{\boldsymbol{\psi}\}]. \tag{4.4}$$

The $\mathcal{B}$-functional defined by Eqs. (2.12) and (2.13) is now decomposed following Eq. (4.2) as

$$\mathcal{B}[t; \{\boldsymbol{\psi}\}] = \sum_{k=1}^{K} \mathcal{B}_k[t; \{\boldsymbol{\psi}\}] + \delta\mathcal{B}[\boldsymbol{\psi}(t)]. \tag{4.5}$$

The memory containing $\mathcal{B}_k$-functional, arising from individual exponential component in Eq. (4.2), is defined via

$$B_k[t; \{\psi\}] \equiv \int_{t_0}^{t} d\tau \, e^{-\gamma_k(t-\tau)} Q[\psi(\tau)], \tag{4.6}$$

as

$$\mathcal{B}_k[t; \{\boldsymbol{\psi}\}] \equiv -i\{c_k B_k[t; \{\boldsymbol{\psi}\}] - c_k^* B_k'[t; \{\psi'\}]\}. \tag{4.7}$$

It satisfies

$$\partial_t \mathcal{B}_k[t; \{\boldsymbol{\psi}\}] = -\gamma_k \mathcal{B}_k[t; \{\boldsymbol{\psi}\}] - i\{c_k Q[\psi(t)] - c_k^* Q[\psi'(t)]\}. \tag{4.8}$$

The $\delta\mathcal{B}$-functional of the last term in Eq. (4.5) contains no memory. It arises from the white-noise residue of bath correlation function in Eq. (4.2).

We obtain

$$\delta\mathcal{B}[\boldsymbol{\psi}(t)] = -i\Delta_N \left(Q[\psi(t)] - Q[\psi'(t)] \right) = -i\Delta_N \mathcal{A}[\boldsymbol{\psi}(t)], \qquad (4.9)$$

with Eq. (2.11) being used in writing the last identity. It leads to a white-noise residue dissipation functional term to Eq. (4.4), i.e.,

$$\delta\mathcal{R}[\boldsymbol{\psi}(t)] = i\mathcal{A}[\boldsymbol{\psi}(t)]\delta\mathcal{B}[\boldsymbol{\psi}(t)] = \Delta_N \mathcal{A}[\boldsymbol{\psi}(t)]\mathcal{A}[\boldsymbol{\psi}(t)], \qquad (4.10)$$

or the superoperator equivalence of $\delta\mathcal{R} = \Delta_N \mathcal{Q}^2$. Here, $\mathcal{Q}$ is defined via $\mathcal{Q}\hat{O} \equiv [\hat{Q}, \hat{O}]$.

HEOM can now be constructed via the auxiliary influence functional,

$$\mathcal{F}_{\mathsf{n}} \equiv \mathcal{F}_{n_1,\ldots,n_K} = \prod_{k=1}^{K} \mathcal{B}_k^{n_k} \mathcal{F}, \qquad (4.11)$$

to the propagator in the ADO $\rho_{\mathsf{n}}(t) = \mathcal{U}_{\mathsf{n}}(t, t_0)\rho(t_0)$ in the HEOM theory; see Eq. (4.13). The ADO's labeling index $\mathsf{n} = \{n_1, \ldots, n_K\}$ specifies the set of non-negative integers involved in Eq. (4.11), with individual n_k for the kth exponential term in Eq. (4.2). Denote also the associated index set of $\mathsf{n}_k^{\pm}$, which differs from n only by changing the specified n_k to $n_k \pm 1$. Carrying out the time derivative on $\mathcal{F}_{\mathsf{n}}$ by using Eq. (4.8) and the identity of $\partial_t\mathcal{F} = -\mathcal{R}\mathcal{F}$, with Eqs. (4.4)–(4.7) and Eq. (4.10), we obtain

$$\partial_t\mathcal{F}_{\mathsf{n}} = -\sum_{k=1}^{K} n_k\gamma_k\mathcal{F}_{\mathsf{n}} - \delta\mathcal{R}\mathcal{F}_{\mathsf{n}} - i\sum_{k=1}^{K} \mathcal{A}\mathcal{F}_{\mathsf{n}_k^+}$$

$$-i\sum_{k=1}^{K} n_k(c_k Q[\psi(t)]\mathcal{F}_{\mathsf{n}_k^-} - c_k^*\mathcal{F}_{\mathsf{n}_k^-} Q[\psi'(t)]). \qquad (4.12)$$

Define now the ADOs via $\rho_{\mathsf{n}}(t) \equiv \mathcal{U}_{\mathsf{n}}(t, t_0)\rho(t_0)$, with the propagators having the path integral expression of

$$\mathcal{U}_{\mathsf{n}}(\boldsymbol{\psi}, t; \boldsymbol{\psi}_0, t_0) = s_{\mathsf{n}} \int_{\boldsymbol{\psi}_0[t_0]}^{\boldsymbol{\psi}[t]} \mathcal{D}\boldsymbol{\psi}\, e^{iS[\psi]} \mathcal{F}_{\mathsf{n}}[\boldsymbol{\psi}] e^{-iS[\psi']}. \qquad (4.13)$$

With the scaling factor of $s_{\mathsf{n}} = \prod_k (n_k!|c_k|^{n_k})^{-1/2}$, the ADOs defined above are not just dimensionless, but also scaled properly to have a uniform error tolerance.[34] Consequently, Eq. (4.12) amounts to the HEOM formalism of

$$
\dot{\rho}_{\mathsf{n}} = -\left[i\mathcal{L}(t) + \sum_{k=1}^{K} n_k \gamma_k + \Delta_N \mathcal{Q}^2 \right] \rho_{\mathsf{n}} - i \sum_{k=1}^{K} \sqrt{(n_k+1)|c_k|}\, \mathcal{Q}\rho_{\mathsf{n}_k^+}
$$

$$
- i \sum_{k=1}^{K} \sqrt{\frac{n_k}{|c_k|}} (c_k \hat{Q} \rho_{\mathsf{n}_k^-} - c_k^* \rho_{\mathsf{n}_k^-} \hat{Q}). \tag{4.14}
$$

Here $\mathcal{Q}\hat{O} \equiv [\hat{Q}, \hat{O}]$. The above HEOM formalism has the generic form of Eq. (2.1), with now the explicit expressions of

$$
\gamma_{\mathsf{n}} = \sum_{k=1}^{K} n_k \gamma_k \quad \text{and} \quad \delta\mathcal{R}_{\mathsf{n}} = \Delta_N \mathcal{Q}^2, \tag{4.15}
$$

together with those for the tier-up and tier-down dependents, $\rho_{\mathsf{n}}^{\{+\}}$ and $\rho_{\mathsf{n}}^{\{-\}}$, by the last two terms in Eq. (4.14), respectively. It should be noted that Eq. (4.14) assumes all exponents $\{\gamma_k\}$ in Eq. (4.2) are real. The HEOM formalism for a general case can be found in Ref. 31, for example, with the BOs-based memory-frequency decomposition of environment correlation function, where the exponents from underdamped BO modes are complex.

It has been shown[31] that for an optimized HEOM theory is concerned, Drude dissipation goes the best with the $[N/N]$ Bose function approximant. Thus, Eq. (4.14) is the optimized HEOM theory for Drude dissipation.[31,35,36] It has also been proved that the $[N+1/N]$-based HEOM construction is the best for BOs dissipation, provided all damping parameters being of $r_{\mathrm{BO}} \geq 0.5$.[31]

4.2. *Accuracy control on white-noise residue ansatz*

The memory-frequency decomposition (i.e., the exponential expansion) of environment correlation function $C(t)$ in Eq. (4.2) dictates the explicit HEOM expressions. In this sense, the adopted memory-frequency decomposition serves as the "statistical bath basis set" of size K for the HEOM construction, up to a sufficiently large level L of tier truncation. The only approximation is then about the treatment of residue outside the basis set

K-space. In comparing between Eq. (4.2) and Eq. (3.1), it is the white-noise-residue ansatz,[13,37]

$$\delta C_N(t) \equiv C(t) - \sum_{k=1}^{K} c_k e^{-\gamma_k t} \approx 2\Delta_N \delta(t). \tag{4.16}$$

It leads to the following residue dissipation superoperator common at all tiers of HEOM [Eq. (2.1)]:

$$\delta\mathcal{R}_n \cdot = \Delta_N[\hat{Q}, [\hat{Q}, \cdot]]. \tag{4.17}$$

This is by far the only generally controllable and also often the best among various tested methods of residue treatment.[13,30,33]

The optimized HEOM theory[31] goes with the minimum basis set of size K and the aforementioned white-noise-residue treatment. The accuracy control can then be analyzed in principle by considering the fact that the only approximation involved in HEOM is Eq. (4.16). The criterion on the applicability of HEOM comprises therefore the conditions under which $\delta C_N(t)$ and its effect on the reduced system dynamics can be treated as Markovian white noise.[30,35,36] We demand *a priori* accuracy control or estimation on the resulting HEOM dynamics for general systems at finite temperatures. Apparently, the unapproximated $\delta C_N(t)$ is a real and even function. The residue spectrum $\delta C_N(\omega) \equiv \frac{1}{2} \int dt\, e^{i\omega t} \delta C_N(t)$ is symmetric.

The validation of white-noise-residue ansatz goes as follows. (i) *Prerequisite of line shape*: Residue spectrum $\delta C_N(\omega)$ be monotonic in $\omega \in [0, \infty)$, varying from $\delta C_N(\omega = 0) = \Delta_N \neq 0$ to $\delta C_N(\omega \to \infty) = 0$; (ii) *Control parameters* $\{\Gamma_N, \kappa_N\}$: Define the residue modulation parameter[30,38,39]:

$$\kappa_N \equiv |\Gamma_N/\Delta_N|^{1/2}, \tag{4.18}$$

with Γ_N being the residue spectrum width, at which $\delta C_N(\omega = \Gamma_N) = \delta C_N(\omega = 0)/2 = \Delta_N/2$. Denote the characteristic system frequency as Ω_s. The white-noise-residue ansatz becomes exact when $\Gamma_N \gg \Omega_s$ and $\kappa_N \gg 1$. In practice it is found that the HEOM dynamics is numerically accurate when[30,35,36]

$$\min\{\Gamma_N/\Omega_s, \kappa_N\} \gtrsim 5. \tag{4.19}$$

It can be used as the accuracy control criterion, upon the residue line shape prerequisite is satisfied. This is an efficient while controllable resum treatment for a partial retrieval of the residue influence, outside the finite decomposition (basis-set) space.[30,31,35,36] The optimized HEOM theory can therefore be established, with *a priori* accuracy control on the only approximation, the white-noise-residue ansatz involved in the HEOM construction.

4.3. *Efficient HEOM propagator: Numerical filtering and indexing algorithm*

The HEOM takes into account the statistical environment bath basis set of K-space size, and it should also be truncated at a finite tier level L, with setting all $\rho_n|_{n>L}$ zero (cf. Eq. (2.1)). Recall that $n_k \geq 0$, with $k = 1, \ldots, K$, the total number of ADOs involved in the HEOM evaluation is then

$$\mathcal{N}(L, K) = \sum_{n=0}^{L} \frac{(n + K - 1)!}{n!(K - 1)!} = \frac{(L + K)!}{L!K!} = \binom{L + K}{K}. \quad (4.20)$$

For later use, we set also $\binom{m < M}{M} \equiv 0$. To locate the ADOs in a numerical HEOM code, we map the multiple indices $n = \{n_1, \ldots, n_K\}$ to an integer $j_n \in [0, \mathcal{N})$, i.e., $\rho_n = \rho_{j_n}$. We choose[40]

$$j_n = \binom{K + n - 1}{K} + \sum_{k=1}^{K} \binom{K + n - k - 1 - \sum_{j=1}^{k} n_j}{K - k}. \quad (4.21)$$

This scheme sorts the ADO-indexes into the tier-based blocks, followed by the sub-indices at the same tier $n = n_1 + \cdots + n_K$. Specifically, we have $j_{n=\{0,\ldots,0\}} = 0$, $j_{n=\{1,0,\ldots,0\}} = 1$, $j_{n=\{0,1,\ldots,0\}} = 2$, and so on.

In principle, an exact HEOM formalism goes with $L_{max} \rightarrow \infty$. However, it converges often at a lower value of L, especially when bath correlation functions are of short memory.[12] It is also noticed that the effect of system-bath coupling on the reduced system density operator dynamics is accounted for at least to the $(2L)$th-order. The above observations conclude that the HEOM construction accounts physically the combined effects of many-particle interaction or anharmonicity of the reduced system, the system-bath coupling strength and memory time. It would be anticipated that the HEOM formalism be of uniform convergence.

The HEOM formalism involves a total number $\mathcal{N}$ of ADOs, and each is a matrix of the reduced system's size. As $\mathcal{N}$ of Eq. (4.20) is concerned, HEOM resembles a "full configuration interaction" construction in quantum chemistry, but refers to the bath configurations that may be termed as "dissipatons" participating in the HEOM construction. The number of ADOs can be huge if large L and K are required. The optimized HEOM theory with a minimum K-construction addresses this issue from the formulation aspect.[20,21,30,31] On the other hand, it is also observed that often there is only a small fraction of total ADOs significant to the reduced system dynamics. Based on this observation,[34] an efficient on-the-fly filtering algorithm is proposed which dramatically reduces the number of involving ADOs during numerical propagation. To validate the filtering algorithm, ADOs are scaled to be dimensionless and to have a uniform error tolerance as that of the primary reduced density operator. Then at each step of propagation, an individual ADO would be considered to be zero if its norm is smaller than the pre-chosen error tolerance. It is found that setting the filtering tolerance to be 2×10^{-5} is sufficient in all cases we have tested by now. Apparently the filtering algorithm also automatically truncates the hierarchy level L, on-the-fly during the numerical propagation.[34]

5. HEOM in Quantum Mechanics for Open Systems

5.1. *The HEOM space and the Schrödinger picture*

We refer the HEOM space to the linear space defined by the HEOM formalism, Eq. (2.1), that can be cast in a matrix-vector form,

$$\dot{\rho}(t) = -i\mathcal{L}(t)\rho(t). \tag{5.1}$$

It defines the HEOM-space dynamics in Schrödinger picture, where the state (column) vector, $\rho \equiv \{\rho_{n=0}; \rho_{n\neq0}\} \equiv \{\rho; \rho_{n\neq0}\}$, consists of all ADOs that are time-dependent. The HEOM-space generator or Liouvillian, $\mathcal{L}(t)$, assumes a matrix form, determined by Eq. (4.14) for Drude dissipation or Eq. (2.1) in general. In contact with the expectation value of an arbitrary reduced system dynamical variable $\hat{A}$, we introduce the HEOM-space inner product, $\langle\langle A|\rho\rangle\rangle$, via the following identities,

$$\bar{A}(t) = \text{tr}[\hat{A}\rho(t)] \equiv \langle\langle\hat{A}|\rho(t)\rangle\rangle = \langle\langle A|\rho(t)\rangle\rangle \equiv \sum_{\text{all } n}\langle\langle\hat{A}_n|\rho_n(t)\rangle\rangle. \quad (5.2)$$

Here, $A = \{\hat{A}_{n=0}; \hat{A}_{n\neq0}\}$, arranged in a row vector, denotes the HEOM-space extension of the system dynamics operator $\hat{A}$. Implied here are also that $\hat{A}_{n=0} \equiv \hat{A}$ and $\hat{A}_{n\neq0} \equiv 0$. They will serve as the initial values to the Heisenberg picture for the HEOM-space evolution of reduced system dynamics operator (see Sec. 5.2).

Discuss now the Dyson equation that is closely related to the interaction picture. To that end, we separate the HEOM-space Liouvillian into two parts, $\mathcal{L}(t) = \mathcal{L}_s + \mathcal{L}'(t)$. Let the unperturbed $\mathcal{L}_s$ be time-independent, while the perturbed $\mathcal{L}'(t)$ be time-dependent in general. Introduce the HEOM-space propagator, $\mathcal{G}(t, \tau)$, that satisfies $\partial_t\mathcal{G}(t, \tau) = -i\mathcal{L}(t)\mathcal{G}(t, \tau)$, by which $\rho(t) = \mathcal{G}(t, \tau)\rho(\tau)$. Denote also $\mathcal{G}_s(t) \equiv \exp(-i\mathcal{L}_s t)$. Elementary linear-space algebra in quantum mechanics leads immediately to the HEOM-space Dyson equation of

$$\mathcal{G}(t, t_0) = \mathcal{G}_s(t - t_0) - i\int_{t_0}^t d\tau\mathcal{G}(t, \tau)\mathcal{L}'(\tau)\mathcal{G}_s(\tau - t_0), \quad (5.3)$$

where t_0 denotes any time before τ at which the specified time-dependent perturbation takes action.

Consider now the perturbed component of reduced system density operator, $\delta\rho(t) = \text{tr}_\text{B}[\rho_\text{total}(t) - \rho_\text{total}^\text{eq}(T)]$, which is just the zeroth-tier ADO in $\delta\rho(t) = \rho(t) - \rho^\text{eq}(T)$. Here $\rho^\text{eq}(T) \equiv \{\rho^\text{eq}(T); \rho_{n\neq0}^\text{eq}(T)\}$ denotes the initial thermal equilibrium HEOM-space state, in the absence of time-dependent perturbation. It can thus be evaluated via the steady-state solution to the HEOM of $\dot{\rho} = -i\mathcal{L}_s\rho = 0$, resulting in not just the reduced system $\rho^\text{eq}(T)$, but also other ADOs, $\rho_{n\neq0}^\text{eq}(T)$, that are nonzero in general, due to the initial system and bath correlation. Apparently, $\mathcal{G}_s(t)\rho^\text{eq}(T) = \rho^\text{eq}(T)$. We obtain therefore

$$\delta\rho(t) = -i\int_{-\infty}^t d\tau\,\mathcal{G}(t, \tau)\mathcal{L}'(\tau)\rho^\text{eq}(T). \quad (5.4)$$

It together with Eq. (5.3) constitute the key results on the interaction picture of HEOM-space dynamics.

Examine now the form of $\mathcal{L}'(t)$, arising from the system Hamiltonian perturbation, $H'(t) = -D\epsilon_{\text{ext}}(t)$, in the presence of time-dependent classical external field. In this case, $\mathcal{L}'(t) = -\mathcal{D}\mathcal{I}\,\epsilon_{\text{ext}}(t)$, as inferred from Eq. (2.1) or Eq. (4.14), is diagonal in the HEOM space, where $\mathcal{I}$ denotes the HEOM-space unit operator, while $\mathcal{D} = \overrightarrow{\mathcal{D}} - \overleftarrow{\mathcal{D}}$ specifies the system operator D-commutator, with $\overrightarrow{\mathcal{D}}\hat{O} = D\hat{O}$ and $\overleftarrow{\mathcal{D}}\hat{O} = \hat{O}D$. We have therefore $\mathcal{L}'(t)\rho = -(\overrightarrow{\mathcal{D}} - \overleftarrow{\mathcal{D}})\rho\,\epsilon_{\text{ext}}(t)$, with

$$\overrightarrow{\mathcal{D}}\rho = \{\hat{\mu}\rho;\ \hat{\mu}\rho_{n\neq 0}\}, \quad \overleftarrow{\mathcal{D}}\rho = \{\rho\hat{\mu};\ \rho_{n\neq 0}\hat{\mu}\}. \tag{5.5}$$

These kicked states are non-Hermite, but can be used individually, in conjunction with Eq. (5.4), as the initial state to the HEOM evaluation of various correlation/response functions of system properties.

5.2. HEOM in the Heisenberg picture

Let us revisit the Schrödinger picture of the HEOM-space dynamics, $\tilde{\rho}(t) = \mathcal{G}_s(t)\tilde{\rho}(0)$, in the absence of time-dependent external field, but with such as the aforementioned non-Hermite initial state $\tilde{\rho}(0) = \overrightarrow{\mathcal{D}}\rho$ or $\overleftarrow{\mathcal{D}}\rho$. Apparently, the HEOM formalism remains valid.

Taking the Drude dissipation as example, Eq. (4.14) is now recast as

$$\dot{\tilde{\rho}}_{\text{n}} = -(i\mathcal{L}_s + \gamma_n + \Delta_N \mathcal{Q}^2)\tilde{\rho}_{\text{n}} - i\sum_{k=1}^{K}\sqrt{(n_k+1)|c_k|}\,\mathcal{Q}\tilde{\rho}_{\text{n}_k^+}$$

$$- i\sum_{k=1}^{K}\sqrt{\frac{n_k}{|c_k|}}(c_k\hat{Q}\tilde{\rho}_{\text{n}_k^-} - c_k^*\tilde{\rho}_{\text{n}_k^-}\hat{Q}). \tag{5.6}$$

To obtain the corresponding Heisenberg picture, let us write down explicitly the key ingredients in the matrix-vector form of Eq. (5.6):

$$\frac{\partial}{\partial t}\begin{bmatrix} \times \\ \tilde{\rho}_{\text{n}_k^-} \\ \tilde{\rho}_{\text{n}} \\ \tilde{\rho}_{\text{n}_k^+} \\ \times \end{bmatrix} = -i\begin{bmatrix} \times & \times & 0 & 0 & 0 \\ \times & \times & \mathcal{B}_{\text{n}_k^-} & 0 & 0 \\ 0 & \mathcal{A}_{\text{n}} & \mathcal{L}_{\text{n}} & \mathcal{B}_{\text{n}} & 0 \\ 0 & 0 & \mathcal{A}_{\text{n}_k^+} & \times & \times \\ 0 & 0 & 0 & \times & \times \end{bmatrix}\begin{bmatrix} \times \\ \tilde{\rho}_{\text{n}_k^-} \\ \tilde{\rho}_{\text{n}} \\ \tilde{\rho}_{\text{n}_k^+} \\ \times \end{bmatrix}. \tag{5.7}$$

Here, $i\mathcal{L}_n$ denotes the quantity inside the first parentheses in Eq. (5.6). While the expressions of $\mathcal{A}_n$ and $\mathcal{B}_n$ can be easily obtained from Eq. (5.6), the associated superoperators $\mathcal{A}_{n_k^+}$ and $\mathcal{B}_{n_k^-}$ can be directly given as $\mathcal{A}_{n_k^+}\hat{O} \equiv \sqrt{\frac{n_k+1}{|c_k|}}\,(c_k Q\hat{O} - c_k^* \hat{O}Q)$ and $\mathcal{B}_{n_k^-}\hat{O} \equiv \sqrt{n_k|c_k|}\,Q\hat{O}$. Both of them then contribute to the Heisenberg picture, see Eq. (5.8). With the matrix $\mathcal{L}_s$ being specified explicitly in Eq. (5.7), the Heisenberg equation of motion, $\dot{A}(t) = -iA(t)\mathcal{L}_s$ for the row vector of $A = \{\times, \hat{A}_{n_k^-}, \hat{A}_n, \hat{A}_{n_k^+}, \times\}$, can be obtained immediately. We have $\dot{\hat{A}}_n = -i\hat{A}_n\mathcal{L}_n - i\hat{A}_{n_k^-}\mathcal{B}_{n_k^-} - i\hat{A}_{n_k^+}\mathcal{A}_{n_k^+}$, following the abbreviated notation as Eq. (5.7). It leads to the final HEOM formalism in the Heisenberg picture, the explicit expression of[41]

$$\dot{\hat{A}}_n(t) = -\hat{A}_n(t)(i\mathcal{L}_s + \gamma_n + \Delta_N Q^2) - i\sum_{k=1}^{K} \sqrt{n_k|c_k|}\,\hat{A}_{n_k^-}(t)Q$$

$$-i\sum_{k=1}^{K} \sqrt{\frac{n_k+1}{|c_k|}}\,[c_k\hat{A}_{n_k^+}(t)\hat{Q} - c_k^*\hat{Q}\hat{A}_{n_k^+}(t)]. \qquad (5.8)$$

Here, $\hat{O}\mathcal{L}_s = [\hat{O}, H_s]$ and $\hat{O}Q = [\hat{O}, \hat{Q}]$, following the identities of $\mathcal{L}_s\hat{O} = [H_s, \hat{O}]$ and $Q\hat{O} = [\hat{Q}, \hat{O}]$ defined earlier. Together with the initial conditions of $\hat{A}_{n=0}(0) = \hat{A}$ and $\hat{A}_{n\neq0}(0) = 0$, the above equation determines the HEOM evolution of $A(t) = A(0)\mathcal{G}_s(t) = A(0)\exp(-i\mathcal{L}_s t)$ for an arbitrary system dynamical variable $\hat{A}$ that can be non-Hermite.

5.3. *Mixed Heisenberg–Schrödinger block-matrix dynamics in nonlinear optical response functions*

To illustrate the versatility of HEOM, let us consider the third-order optical response functions, as probed by coherent two-dimensional spectroscopies, operated with short pulsed fields in certain four-wave-mixing configurations.[42,43] Following the standard interaction picture algebra, but now with Eq. (5.4), we obtain the third-order optical response function in HEOM space as

$$R^{(3)}(t_3, t_2, t_1) = \langle\!\langle \mu_{k_s}|\mathcal{G}(t_3)\mathcal{D}_{k_3}\mathcal{G}(t_2)\mathcal{D}_{k_2}\mathcal{G}(t_1)\mathcal{D}_{k_1}|\rho^{eq}\rangle\!\rangle. \qquad (5.9)$$

It is just the HEOM space analogue of the conventional full system-plus-bath material space expression.[44,45] Apparently the mixed Heisenberg–Schrödinger scheme will greatly facilitate the evaluation of third-order response function.

As a kind of four-wave-mixing spectroscopy, the wavevector $\mathbf{k}_s$ of the signal field satisfies the phase-matching condition. It is that $\mathbf{k}_s = \pm\mathbf{k}_3 \pm \mathbf{k}_2 \pm \mathbf{k}_1$, in relation to the three incident pulsed fields interacting with the system sequentially. Three basic configurations of coherent two-dimensional spectroscopy[42,43] will be classified in Sec. 5.3. Their efficient evaluation via block-HEOM in mixed Heisenberg–Schrödinger scheme will be detailed in the following part.

To begin with, we show that the optical response function can be recast in the block-HEOM dynamics form. Consider the third-order optical processes involving the initial ground $|g\rangle$, the excited $|e\rangle$, and doubly-excited $|f\rangle$ manifolds of electronic states. Assume also that the relaxation between different manifolds is negligible. This implies that not only the system Hamiltonian but also each dissipative mode $\hat{Q}$ is block diagonalized, in virtue of Born–Oppenheimer principle.

On the other hand, the transition dipole operators involved in the third-order optical response function are also in the block-matrix form, although not diagonal. For example, the transition dipole $\mu_- \equiv \hat{\mu}_{ge}|g\rangle\langle e|+\hat{\mu}_{ef}|e\rangle\langle f|$ amounts explicitly to

$$\mu_- = \begin{bmatrix} \hat{O}_{gg} & \hat{\mu}_{ge} & \hat{O}_{gf} \\ \hat{O}_{eg} & \hat{O}_{ee} & \hat{\mu}_{ef} \\ \hat{O}_{fg} & \hat{O}_{fe} & \hat{O}_{ff} \end{bmatrix}. \tag{5.10}$$

The matrix $\hat{O}_{uv}$ in the (uv)-block, with $u, v \in \{g, e, f\}$, has the order of $N_u \times N_v$. We have $N_g = 1$, $N_e = M$ and $N_f = M(M-1)/2$, for a molecular aggregate of size M, assuming the simple exciton model.

Expand the third-order optical response function of Eq. (5.9) in the eight Liouville-space pathways, as depicted in upper part of Fig. 2, and their complex conjugate contributions.[44,45] We obtain

$$R^{(3)}(t_3, t_2, t_1) = i^3 \sum_{\alpha=1}^{8} [R_\alpha(t_3, t_2, t_1) - \text{c.c.}], \tag{5.11}$$

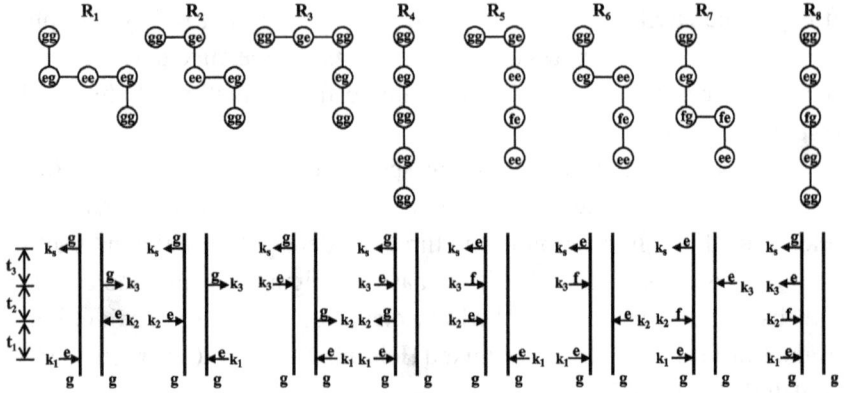

Fig. 2. Eight Liouville-space pathways (upper) and double-sided Feynman diagrams (lower) for two-dimensional spectroscopy in the rotating-wave approximation. Each Liouville-space pathway starts from the upper-left circle, while the double-sided Feynman diagram starts from the bottom, following the convention of Ref. 45.

with[41]

$$R_1(t_3, t_2, t_1) = \langle\langle\mu_{ge}|\mathcal{G}_{eg}(t_3)\overleftarrow{\mu}_{eg}\mathcal{G}_{ee}(t_2)\overrightarrow{\mu}_{ge}\mathcal{G}_{eg}(t_1)\overrightarrow{\mu}_{eg}|\rho_{gg}^{eq}\rangle\rangle$$
$$\times e^{-i\omega_{eg}(t_3+t_1)},$$

$$R_2(t_3, t_2, t_1) = \langle\langle\mu_{ge}|\mathcal{G}_{eg}(t_3)\overleftarrow{\mu}_{eg}\mathcal{G}_{ee}(t_2)\overrightarrow{\mu}_{eg}\mathcal{G}_{ge}(t_1)\overleftarrow{\mu}_{ge}|\rho_{gg}^{eq}\rangle\rangle$$
$$\times e^{-i\omega_{eg}(t_3-t_1)},$$

$$R_3(t_3, t_2, t_1) = \langle\langle\mu_{ge}|\mathcal{G}_{eg}(t_3)\overrightarrow{\mu}_{eg}\mathcal{G}_{gg}(t_2)\overleftarrow{\mu}_{eg}\mathcal{G}_{ge}(t_1)\overleftarrow{\mu}_{ge}|\rho_{gg}^{eq}\rangle\rangle$$
$$\times e^{-i\omega_{eg}(t_3-t_1)},$$

$$R_4(t_3, t_2, t_1) = \langle\langle\mu_{ge}|\mathcal{G}_{eg}(t_3)\overrightarrow{\mu}_{eg}\mathcal{G}_{gg}(t_2)\overrightarrow{\mu}_{ge}\mathcal{G}_{eg}(t_1)\overrightarrow{\mu}_{eg}|\rho_{gg}^{eq}\rangle\rangle$$
$$\times e^{-i\omega_{eg}(t_3+t_1)},$$

$$R_5(t_3, t_2, t_1) = -\langle\langle\mu_{ef}|\mathcal{G}_{fe}(t_3)\overrightarrow{\mu}_{fe}\mathcal{G}_{ee}(t_2)\overrightarrow{\mu}_{eg}\mathcal{G}_{ge}(t_1)\overleftarrow{\mu}_{ge}|\rho_{gg}^{eq}\rangle\rangle$$
$$\times e^{-i(\omega_{fe}t_3-\omega_{eg}t_1)},$$

$$R_6(t_3, t_2, t_1) = -\langle\langle\mu_{ef}|\mathcal{G}_{fe}(t_3)\overleftarrow{\mu}_{fe}\mathcal{G}_{ee}(t_2)\overrightarrow{\mu}_{ge}\mathcal{G}_{eg}(t_1)\overrightarrow{\mu}_{eg}|\rho_{gg}^{eq}\rangle\rangle$$
$$\times e^{-i(\omega_{fe}t_3+\omega_{eg}t_1)},$$

$$R_7(t_3, t_2, t_1) = -\langle\langle\mu_{ef}|\mathcal{G}_{fe}(t_3)\overleftarrow{\mu}_{ge}\mathcal{G}_{fg}(t_2)\overrightarrow{\mu}_{fe}\mathcal{G}_{eg}(t_1)\overrightarrow{\mu}_{eg}|\rho_{gg}^{eq}\rangle\rangle$$
$$\times e^{-i(\omega_{fe}t_3+\omega_{fg}t_2+\omega_{eg}t_1)},$$

$$R_8(t_3, t_2, t_1) = \langle\!\langle \mu_{ge} | \mathcal{G}_{eg}(t_3) \overrightarrow{\mu}_{ef} \mathcal{G}_{fg}(t_2) \overrightarrow{\mu}_{fe} \mathcal{G}_{eg}(t_1) \overrightarrow{\mu}_{eg} | \rho_{gg}^{eq} \rangle\!\rangle,$$
$$\times e^{-i(\omega_{eg}t_3 + \omega_{fg}t_2 + \omega_{eg}t_1)}.$$

$$(5.12)$$

The underlying optical processes will be discussed in Sec. 6. The electronic phase factor in each individual R_α can be formally absorbed into the involving Green's functions, i.e., $\mathcal{G}_{uv}(t)e^{-i\omega_{uv}t} \to \mathcal{G}_{uv}(t)$, with $\omega_{uv} \equiv \epsilon_u - \epsilon_v$ denoting the chosen reference frequency for the optical transition between two specified electronic manifolds. In present notion, $\mathcal{G}_{uv}(t)$ involves only slow motion dynamics, as the highly oscillatory optical frequency component is factorized out for numerical advantage. The corresponding block HEOM dynamics in both Schrödinger and Heisenberg pictures will be detailed later in this subsection. There are other advantages for the present notion. The overall electronic phase factor can in fact be used to distinguish rephasing versus non-rephasing optical processes[44–46] and also to visualize the rotating wave approximation as seen below.

To implement the third-order optical response functions in Eq. (5.12), we start with the thermal equilibrium ρ_{gg}^{eq} in the ground-state $|g\rangle$-manifold. As described earlier, it is determined by the steady state solution to HEOM, involving now only the (gg)-block part. For the simple exciton model, the $|g\rangle$-manifold contains only one level, and the ADOs in (gg)-block are all 1×1 matrices, resulting in $\rho_{gg}^{eq} = \{\rho_{n=0}^{gg,eq} = 1, \rho_{n\neq0}^{gg,eq} = 0\}$. Block-matrix multiplications are then followed:

$$\overrightarrow{\mu}_{eg}\rho_{gg}^{eq} = \{\hat{\mu}_{eg}\rho_{n=0}^{gg,eq}, \hat{\mu}_{eg}\rho_{n\neq0}^{gg,eq}\} \equiv \tilde{\rho}_{eg}(0),$$
$$\overleftarrow{\mu}_{ge}\rho_{gg}^{eq} = \{\rho_{n=0}^{gg,eq}\hat{\mu}_{ge}, \rho_{n\neq0}^{gg,eq}\hat{\mu}_{ge}\} \equiv \tilde{\rho}_{ge}(0) = \tilde{\rho}_{eg}^{\dagger}(0).$$

$$(5.13)$$

They are the initial states for the block-HEOM $\mathcal{G}_{uv}(t_1)$ propagations in Eq. (5.12). Each ADO within $\tilde{\rho}_{uv}$ is an $N_u \times N_v$ matrix, and Hermite conjugate with its counterpart in $\tilde{\rho}_{vu}$; i.e., $\tilde{\rho}_{vu} = \{\tilde{\rho}_n^{vu}\} = \{(\tilde{\rho}_n^{uv})^{\dagger}\} \equiv \tilde{\rho}_{uv}^{\dagger}$.

The t_1- and t_2-propagations in Eq. (5.12) are implemented in a nested manner in Schrödinger picture. This picture is defined via the action-from-left of $\mathcal{G}_{uv}(t)$ on state variables; e.g., $\tilde{\rho}_{uv}(t) = \mathcal{G}_{uv}(t)\tilde{\rho}_{uv}(0)$.

The block-HEOM in Schrödinger picture can be reduced from Eq. (5.6) as

$$\dot{\tilde{\rho}}_n^{uv} = -i(\mathcal{L}_{uv} + \gamma_n + \Delta_N \mathcal{Q}_{uv}^2)\tilde{\rho}_n^{uv}$$

$$- i \sum_{k=1}^{K} \sqrt{(n_k + 1)|c_k|}\, \mathcal{Q}_{uv}\tilde{\rho}_{n_k^+}^{uv}$$

$$- i \sum_{k=1}^{K} \sqrt{\frac{n_k}{|c_k|}}\, (c_k Q_{uu}\tilde{\rho}_{n_k^-}^{uv} - c_k^*\tilde{\rho}_{n_k^-}^{uv} Q_{vv}). \tag{5.14}$$

Here, $\mathcal{L}_{uv}\hat{O}_{uv} \equiv H_{uu}\hat{O}_{uv} - \hat{O}_{uv}H_{vv}$ and $\mathcal{Q}_{uv}\hat{O}_{uv} \equiv Q_{uu}\hat{O}_{uv} - \hat{O}_{uv}Q_{vv}$. Equivalently, $\hat{O}_{vu}\mathcal{L}_{uv} = \hat{O}_{vu}H_{uu} - H_{vv}\hat{O}_{vu}$ and $\hat{O}_{vu}\mathcal{Q}_{uv} = \hat{O}_{vu}Q_{uu} - Q_{vv}\hat{O}_{vu}$, which will be used in the following Heisenberg picture.

The t_3-propagation in Eq. (5.12) is implemented in Heisenberg picture, in parallel with the t_1- and t_2-propagations. The Heisenberg picture is defined via the action-from-right of $\mathcal{G}_{uv}(t)$ on dynamic variables, i.e., $A_{vu}(t) = A_{vu}\mathcal{G}_{uv}(t)$. The initial condition is $A_{vu}(0) = A_{vu} = \{A_{n=0}^{vu} = A_{vu}, A_{n\neq0}^{vu} = 0\}$ as discussed before. In consistent with Eq. (5.14) or Eq. (5.8), the block-HEOM in Heisenberg picture reads

$$\dot{A}_n^{vu} = -iA_n^{vu}(\mathcal{L}_{uv} + \gamma_n + \Delta_N \mathcal{Q}_{uv}^2) - i \sum_{k=1}^{K} \sqrt{n_k|c_k|}\, A_{n_k^-}^{vu} \mathcal{Q}_{uv}$$

$$- i \sum_{k=1}^{K} \sqrt{\frac{n_k + 1}{|c_k|}}\, (c_k A_{n_k^+}^{vu} Q_{uu} - c_k^* Q_{vv} A_{n_k^+}^{vu}). \tag{5.15}$$

To evaluate the third-order optical response functions in Eq. (5.12) with block-HEOM in mixed Heisenberg–Schrödinger scheme, we introduce

$$\tilde{\rho}_{uv}(t_2; t_1) \equiv \mathcal{G}_{uv}(t_2)\tilde{\rho}_{uv}(0; t_1), \tag{5.16}$$

for three types of initial t_2 conditions:

$$\tilde{\rho}_{ee}(0; t_1) = \overleftarrow{\mu}_{ge}\tilde{\rho}_{eg}(t_1),$$

$$\tilde{\rho}_{gg}(0; t_1) = \overrightarrow{\mu}_{ge}\tilde{\rho}_{eg}(t_1), \tag{5.17}$$

$$\tilde{\rho}_{fg}(0; t_1) = \overrightarrow{\mu}_{fe}\tilde{\rho}_{eg}(t_1).$$

We can recast Eq. (5.12) as

$$R_1(t_3, t_2, t_1) = \langle\langle \mu_{ge}(t_3) | \overleftarrow{\mu}_{eg} \tilde{\rho}_{ee}(t_2; t_1) \rangle\rangle \, e^{-i\omega_{eg}(t_3+t_1)},$$

$$R_2(t_3, t_2, t_1) = \langle\langle \mu_{ge}(t_3) | \overleftarrow{\mu}_{eg} \tilde{\rho}^{\dagger}_{ee}(t_2; t_1) \rangle\rangle \, e^{-i\omega_{eg}(t_3-t_1)},$$

$$R_3(t_3, t_2, t_1) = \langle\langle \mu_{ge}(t_3) | \overrightarrow{\mu}_{eg} \tilde{\rho}^{\dagger}_{gg}(t_2; t_1) \rangle\rangle \, e^{-i\omega_{eg}(t_3-t_1)},$$

$$R_4(t_3, t_2, t_1) = \langle\langle \mu_{ge}(t_3) | \overrightarrow{\mu}_{eg} \tilde{\rho}_{gg}(t_2; t_1) \rangle\rangle \, e^{-i\omega_{eg}(t_3+t_1)},$$

$$R_5(t_3, t_2, t_1) = -\langle\langle \mu_{ef}(t_3) | \overrightarrow{\mu}_{fe} \tilde{\rho}^{\dagger}_{ee}(t_2; t_1) \rangle\rangle \, e^{-i(\omega_{fe}t_3 - \omega_{eg}t_1)}, \qquad (5.18)$$

$$R_6(t_3, t_2, t_1) = -\langle\langle \mu_{ef}(t_3) | \overrightarrow{\mu}_{fe} \tilde{\rho}_{ee}(t_2; t_1) \rangle\rangle \, e^{-i(\omega_{fe}t_3 + \omega_{eg}t_1)},$$

$$R_7(t_3, t_2, t_1) = -\langle\langle \mu_{ef}(t_3) | \overrightarrow{\mu}_{ge} \tilde{\rho}_{fg}(t_2; t_1) \rangle\rangle \, e^{-i(\omega_{fe}t_3 + \omega_{fg}t_2 + \omega_{eg}t_1)},$$

$$R_8(t_3, t_2, t_1) = \langle\langle \mu_{ge}(t_3) | \overrightarrow{\mu}_{ef} \tilde{\rho}_{fg}(t_2; t_1) \rangle\rangle \, e^{-i(\omega_{eg}t_3 + \omega_{fg}t_2 + \omega_{eg}t_1)}.$$

These are the final expressions for the mixed Heisenberg–Schrödinger scheme block-HEOM evaluation of third-order optical response functions.[41] Note that the numerical filtering algorithm[34] remains valid in both the Schrödinger and Heisenberg pictures of HEOM propagation.

6. Two-Dimensional Spectroscopy: Model Calculations

There are three basic configurations of coherent two-dimensional spectroscopy, and their signals are denoted as $S_{\mathbf{k}_I}$, $S_{\mathbf{k}_{II}}$, and $S_{\mathbf{k}_{III}}$, respectively.[43] For simplicity we adopt the rotating-wave approximation and the impulsive fields limit.

The $S_{\mathbf{k}_I}$ signal goes with $\mathbf{k}_s = \mathbf{k}_3 + \mathbf{k}_2 - \mathbf{k}_1 \equiv \mathbf{k}_I$, the stimulated photon echo or rephasing configuration, while the $S_{\mathbf{k}_{II}}$ signal goes with $\mathbf{k}_s = \mathbf{k}_3 - \mathbf{k}_2 + \mathbf{k}_1 \equiv \mathbf{k}_{II}$ and is non-rephasing. With the aid of the double-sided Feynman diagrams in Fig. 2, these two signals are identified to be

$$S_{\mathbf{k}_{I/II}}(\omega_3, t_2, \omega_1) = \text{Re} \int_0^\infty dt_3 \int_0^\infty dt_1 e^{i(\omega_3 t_3 \mp \omega_1 t_1)} R_{\mathbf{k}_{I/II}}(t_3, t_2, t_1) \quad (6.1)$$

and related respectively to

$$\begin{aligned} R_{\mathbf{k}_I} &= R_2 + R_3 + R_5 \qquad \text{(rephasing)}, \\ R_{\mathbf{k}_{II}} &= R_1 + R_4 + R_6 \qquad \text{(non-rephasing)}. \end{aligned} \qquad (6.2)$$

The rephasing versus non-rephasing nature of individual R_α in Eq. (5.12) or Eq. (6.2) can be inferred easily from its overall electronic phase factor.[44-46] The signs associating with the frequencies ω_3 and ω_1 in Eq. (6.1) are resulted from the incident fields in the specified four-wave-mixing configuration in the impulsive limit. These signs are just opposite to those in the electronic phase factor of participating R_α contributions. Thus, the participated pathway contributions via the electronic rotating wave approximation are also evident in Eq. (6.1).

Experiments can also be performed in the configuration that the pulsed $\mathbf{k}_2$-field is applied continuously not only after but also before the $\mathbf{k}_1$-field. The resulting signal amounts to $S_{\mathbf{k}_\mathrm{I}+\mathbf{k}_\mathrm{II}} = S_{\mathbf{k}_\mathrm{I}} + S_{\mathbf{k}_\mathrm{II}}$. It is in fact the pump-probe absorption configuration, involving all the six pathways R_1 to R_6 contributions. As inferred from Eq. (5.12), these six pathways can be classified into the *excited-state emission* (R_1, R_2), *ground-state bleaching* (R_3, R_4), and *excited-state absorption* (R_5, R_6) contributions. In fact, the t_1 and t_3 represent the excitation and detection time periods, and therefore, the ω_1 and ω_3 in Eq. (6.1) are the excitation and detection frequencies, respectively. The t_2 denotes the waiting time, during which the system is either in the excited or the ground state manifold, with underlying dynamics being governed by $\mathcal{G}_{ee}(t_2)$ or $\mathcal{G}_{gg}(t_2)$, respectively; see Eq. (5.12). The advanced HEOM evaluations of the $S_{\mathbf{k}_\mathrm{I}}$ and $S_{\mathbf{k}_\mathrm{II}}$ signals will be exemplified later, see Figs. 3(a) and 3(b), respectively, for a model system that consists of a total of seven states.

The $S_{\mathbf{k}_\mathrm{III}}$ signal goes with $\mathbf{k}_s = -\mathbf{k}_3 + \mathbf{k}_2 + \mathbf{k}_1 \equiv \mathbf{k}_\mathrm{III}$ and is related to

$$R_{\mathbf{k}_\mathrm{III}} = R_7 + R_8 \quad \text{(double-excitation)}. \tag{6.3}$$

As inferred from the double-sided Feynman diagrams in Fig. 2, the R_7 and R_8 are the *double-excitation absorption* pathways, involving the $|f\rangle \leftarrow |e\rangle \leftarrow |g\rangle$ processes, while the bra state remains in $\langle g|$. During the time t_2 period, the $\mathbf{k}_\mathrm{III}$-configuration explores therefore double quantum coherence dynamics governed by $\mathcal{G}_{fg}(t_2)$, in contrast to the $\mathbf{k}_{\mathrm{I/II}}$-scheme involving electronic state population dynamics. The detection $\mathbf{k}_3$ field involves single-excitation absorption of $\langle e| \leftarrow \langle g|$ in R_7 and single-excitation emission $|e\rangle \leftarrow |f\rangle$ in R_8, as evident in the corresponding double-sided Feynman diagrams, see Ref. 41. The above analysis justifies the fact that the $\mathbf{k}_\mathrm{III}$-signal is designed to probe the correlation between single and double

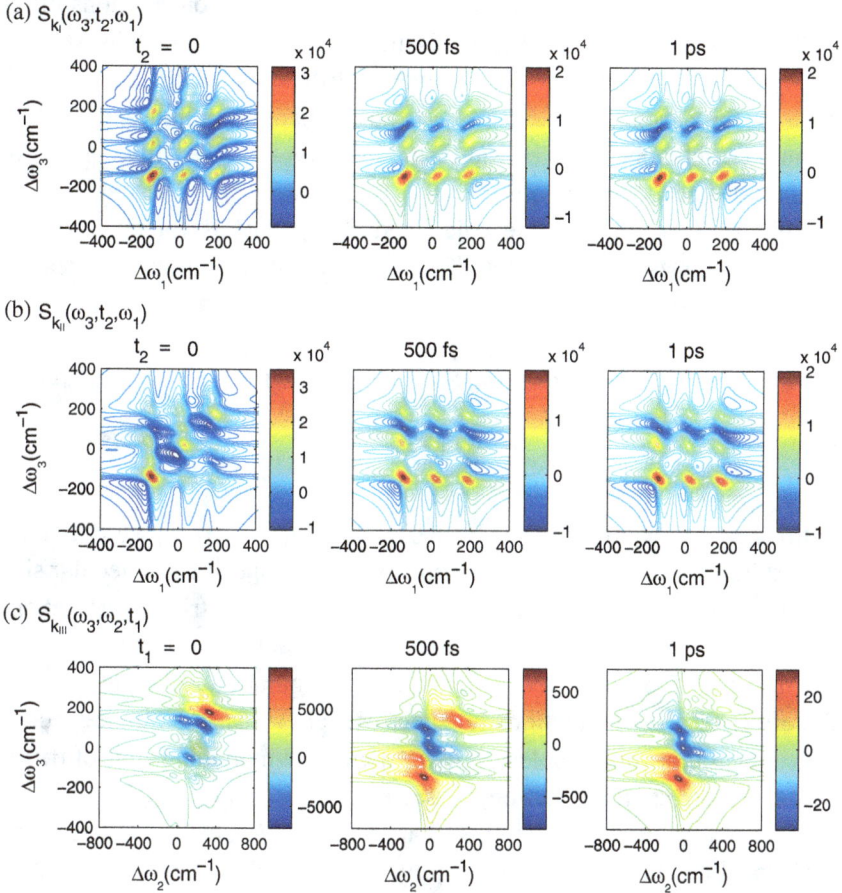

Fig. 3. Coherent two-dimensional spectroscopies for a model excitonic trimer system, at 77 K. See text for details.

excitations.[47–49] The two-dimensional half-Fourier transforms are therefore performed with t_2 to resolve the double-excitation frequency and with either t_3 or t_1 to resolve the specified single-excitation frequency. In this the following demonstration, see Fig. 3(c), we choose

$$S_{\mathbf{k}_{\mathrm{III}}}(\omega_3, \omega_2, t_1) = \mathrm{Re} \int_0^\infty dt_3 \int_0^\infty dt_2 e^{i(\omega_3 t_3 + \omega_2 t_2)} R_{\mathbf{k}_{\mathrm{III}}}(t_3, t_2, t_1). \quad (6.4)$$

In the independent exciton limit, the $\mathbf{k}_{\mathrm{III}}$-signal vanishes. This can be seen from the involving R_7 and R_8 contributions that are associated with *overall opposite signs* [cf. Eq. (5.12)]. Besides that, these two contributions differ by their single coherence dynamics in the t_3 period, which is $\mathcal{G}_{fe}(t_3)$ in R_7 but $\mathcal{G}_{eg}(t_3)$ in R_8. These two contributions would cancel each other in the absence of both inter-exciton transfer coupling and double-exciton correlation. Thus the $\mathbf{k}_{III}$ technique serves as a sensitive probe for interactions between excitons.

Figure 3 exemplifies the calculated $S_{\mathbf{k}_{I/II/III}}$-signals for an excitonic trimer system, as detailed below. The reduced system Hamiltonian reads

$$H = \sum_{m=1}^{3} \epsilon_m \hat{b}_m^\dagger \hat{b}_m + V \sum_{m=1,2} (\hat{b}_m^\dagger \hat{b}_{m+1} + \hat{b}_{m+1}^\dagger \hat{b}_m) + U \sum_{m>n} \hat{b}_m^\dagger \hat{b}_m \hat{b}_n^\dagger \hat{b}_n,$$

(6.5)

with $\epsilon_{m+1} - \epsilon_m = 50\,\text{cm}^{-1}$ and $V = U = 100\,\text{cm}^{-1}$. Here, $\hat{b}_m^\dagger$ $(\hat{b}_m)$ denotes the exciton creation (annihilation) operator on the specified molecular site. Involved in calculations are a total of seven states of the trimer system: one in the ground $|g\rangle$, three in single-exciton $|e\rangle$, and three in double-exciton $|f\rangle$ manifold. Co-linear field polarization configuration is adopted, so that the effect of dipole directions on spectroscopic signals can be neglected. The optical transition dipoles, $\mu_{mx}(\hat{b}_m^\dagger + \hat{b}_m)$, on individual sites of trimer along the polarization direction, are set to be of $\mu_{2x}/\mu_{1x} = \mu_{3x}/\mu_{1x} = 0.2$. The dissipative modes used in Fig. 3 are $\hat{Q}_m = \hat{b}_m^\dagger \hat{b}_m$, $m = 1, 2$ and 3, and each local exciton energy fluctuation is modulated by Drude dissipation [Eq. (4.1)] of $\lambda = 35\,\text{cm}^{-1}$ and $\gamma_D^{-1} = 100\,\text{fs}$, at temperature 77 K.

We have thus specified all system-and-bath details in relation to the HEOM evaluation of coherent two-dimensional spectroscopic signals in Fig. 3. For simplicity, we do not consider correlated fluctuations that can be included in the HEOM evaluation without difficulty. Frequencies are reported in terms of detunings, $\Delta\omega_{1,3} = \omega_{1,3} - \omega_{eg}$, from the reference excitonic transition frequency of $\omega_{eg} = \omega_{fe}$ that is set to be the on-site excitonic energy of $\epsilon_{m=2}$. Similarly, $\Delta\omega_2 = \omega_2 - \omega_{fg}$ is the double-excitation detuning and used in labeling $S_{\mathbf{k}_{III}}$ signals in Fig. 3(c). Static disorders are neglected, as they are irrelevant to the HEOM methodology. The resulting signal peaks in each panel of Fig. 3 are well separated and, therefore, easily characterized with the corresponding excitonic transitions.

They are found to be the localized excitons for the nondegenerate system in study.

The results reported in Fig. 3 are numerically exact, evaluated in an optimally efficient manner and also with *a priori* accuracy control. These are the features of the optimized HEOM methodology detailed in this chapter. As mentioned in Sec. 4, the optimal HEOM construction for Drude dissipation goes with the $[N/N]$-PSD scheme, Eq. (3.7). The accuracy control stipulated in Eqs. (4.18) and (4.19) can be determined readily via the expressions of $\Gamma_N = [r_N + \sqrt{(\beta \gamma_D)^2 + 0.34 r_N^2}]/\beta$ and $\kappa_N = \sqrt{r_N \Gamma_N / \beta \lambda \gamma_D}$, where $r_N = (2N + 2)(2N + 3)$.[30] The quantitative accuracy criterion of Eq. (4.19), $\min\{\Gamma_N / \Omega_s, \kappa_N\} \gtrsim 5$, where Ω_s is the characteristic frequency of the reduced system, leads then to the optimal HEOM construction with the smallest $[N/N]$ scheme. For the system studied in Fig. 3 at $T = 77$ K, the $[1/1]$-PSD scheme is sufficient. We set $L = 20$ for the HEOM truncation level and the standard value of 2×10^{-5} for the filtering error tolerance. The optimized HEOM theory goes also with the mixed Heisenberg–Schrödinger picture and block-matrix implementation of nonlinear optical response functions, as detailed in Sec. 5. The HEOM evaluation of coherent two-dimensional spectroscopy in Fig. 3 takes about a couple minutes on a single processor of Intel(R) Core(TM)2 Q9650 @3.00 GHz.

7. Concluding Remarks

In this chapter, we present a comprehensive account on the recently developed optimized HEOM theory. It is not just a formally exact quantum dissipation theory, but also numerically the most implementable by far. It is capable of addressing transient dynamics of an arbitrary open system, at any finite temperature and in contact with Gaussian-statistics bath environment. The HEOM approach describes quantitatively the energy relaxation and dephasing processes, and also entanglement between system and environment. It resolves nonperturbatively the combined effects of system-bath dissipative couplings, many-particle interactions, and non-Markovian memory. Therefore, the HEOM method has provided a universal, reliable and versatile theoretical tool to investigate quantum dissipation problems at the quantitative level, which involve many important fields at the frontiers of physics and chemistry in general.

Optimized HEOM theory maximizes the range of practical applications of this exact method. It not only acquires the minimum number of dynamics variables (i.e., the ADOs), the optimized theory goes also by *a priori* accuracy control over its numerical application to any given quantum dissipative system at a finite temperature. Optimized HEOM methodology includes also the mixed Heisenberg–Schrödinger scheme and block-matrix implementation, leading to such as an efficient evaluation of third-order optical response function and coherent two-dimensional spectroscopy. Recently, there are also intensive activities in implementing HEOM on parallel computer clusters and graphics processing units.[17, 18]

Acknowledgments

Support from the Hong Kong RGC (605012) and UGC (AoE/P-04/08-2), the NNSF China (21033008 & 21073169), and the National Basic Research Program of China (2010CB923300 & 2011CB921400) is gratefully acknowledged.

References

1. G. S. Engel, T. R. Calhoun, E. L. Read, T. K. Ahn, T. Mančal, Y. C. Cheng, R. E. Blankenship and G. R. Fleming, *Nature* **446**, 782 (2007).
2. H. Lee, Y.-C. Cheng and G. R. Fleming, *Science* **316**, 1462 (2007).
3. T. R. Calhoun, N. S. Ginsberg, G. S. Schlau-Cohen, Y.-C. Cheng, M. Ballottari, R. Bassi and G. R. Fleming, *J. Phys. Chem. B* **113**, 16291 (2009).
4. G. Panitchayangkoon, D. Hayes, K. A. Fransted, J. R. Caram, E. Harel, J. Wen, R. E. Blankenship and G. S. Engel, *Proc. Natl. Acad. Sci. USA* **107**, 12766 (2010).
5. E. Collini, C. Y. Wong, K. E. Wilk, P. M. G. Curmi, P. Brumer and G. D. Scholes, *Nature* **463**, 644 (2010).
6. R. E. Blankenship, *Molecular Mechanisms of Photosynthesis* (Blackwell Science, Oxford, 2002).
7. H. van Amerongen, L. Valkunas and R. van Grondelle, *Photosythetic Exictons* (World Scientific, Singapore, 2000).
8. Y. J. Yan and R. X. Xu, *Annu. Rev. Phys. Chem.* **56**, 187 (2005).
9. Y. Tanimura, *Phys. Rev. A* **41**, 6676 (1990).
10. Y. Tanimura, *J. Phys. Soc. Jpn.* **75**, 082001 (2006).
11. Y. A. Yan, F. Yang, Y. Liu and J. S. Shao, *Chem. Phys. Lett.* **395**, 216 (2004).
12. R. X. Xu, P. Cui, X. Q. Li, Y. Mo and Y. J. Yan, *J. Chem. Phys.* **122**, 041103 (2005).
13. R. X. Xu and Y. J. Yan, *Phys. Rev. E* **75**, 031107 (2007).
14. J. S. Jin, X. Zheng and Y. J. Yan, *J. Chem. Phys.* **128**, 234703 (2008).
15. L. P. Chen, R. H. Zheng, Y. Y. Jing and Q. Shi, *J. Chem. Phys.* **134**, 194508 (2011).

16. J. Zhu, S. Kais, P. Rebentrost and A. Aspuru-Guzik, *J. Phys. Chem. B* **115**, 1531 (2011).
17. C. Kreisbeck, T. Kramer, M. Rodríguez and B. Hein, *J. Chem. Theory Comput.* **7**, 2166 (2011).
18. B. Hein, C. Kreisbeck, T. Kramer and M. Rodríguez, *New J. Phys.* **14**, 023018 (2012).
19. J. Strümpfer and K. Schulten, *J. Chem. Theory Comput.* **8**, 2808 (2012).
20. J. Hu, R. X. Xu and Y. J. Yan, *J. Chem. Phys.* **133**, 101106 (2010).
21. J. Hu, M. Luo, F. Jiang, R. X. Xu and Y. J. Yan, *J. Chem. Phys.* **134**, 244106 (2011).
22. U. Weiss, *Quantum Dissipative Systems*, 3rd ed. Series in Modern Condensed Matter Physics, Vol. 13 (World Scientific, Singapore, 2008).
23. A. O. Caldeira and A. J. Leggett, *Ann. Phys.* **149**, 374 (1983); Erratum **153**, 445 (1984).
24. R. P. Feynman and F. L. Vernon, Jr., *Ann. Phys.* **24**, 118 (1963).
25. X. Zheng, R. X. Xu, J. Xu, J. S. Jin, J. Hu and Y. J. Yan, *Prog. Chem.* **24**, 1129 (2012).
26. N. Makri, *J. Math. Phys.* **36**, 2430 (1995).
27. N. Makri, *J. Phys. Chem. A* **102**, 4414 (1998).
28. R. X. Xu and Y. J. Yan, *J. Chem. Phys.* **116**, 9196 (2002).
29. C. Meier and D. J. Tannor, *J. Chem. Phys.* **111**, 3365 (1999).
30. J. J. Ding, J. Xu, J. Hu, R. X. Xu and Y. J. Yan, *J. Chem. Phys.* **135**, 164107 (2011).
31. J. J. Ding, R. X. Xu and Y. J. Yan, *J. Chem. Phys.* **136**, 224103 (2012).
32. G. A. Baker Jr. and P. Graves-Morris, *Padé Approximants*, 2nd edn. (Cambridge University Press, New York, 1996).
33. K. B. Zhu, R. X. Xu, H. Y. Zhang, J. Hu and Y. J. Yan, *J. Phys. Chem. B* **115**, 5678 (2011).
34. Q. Shi, L. P. Chen, G. J. Nan, R. X. Xu and Y. J. Yan, *J. Chem. Phys.* **130**, 084105 (2009).
35. R. X. Xu, B. L. Tian, J. Xu, Q. Shi and Y. J. Yan, *J. Chem. Phys.* **131**, 214111 (2009).
36. B. L. Tian, J. J. Ding, R. X. Xu and Y. J. Yan, *J. Chem. Phys.* **133**, 114112 (2010).
37. A. Ishizaki and Y. Tanimura, *J. Phys. Soc. Jpn.* **74**, 3131 (2005).
38. R. Kubo, *J. Math. Phys.* **4**, 174 (1963).
39. R. Kubo, *Adv. Chem. Phys.* **15**, 101 (1969).
40. J. Xu, R. X. Xu and Y. J. Yan, *New J. Phys.* **11**, 105037 (2009).
41. J. Xu, R. X. Xu, D. Abramavicius, H. D. Zhang and Y. J. Yan, *Chin. J. Chem. Phys.* **24**, 497 (2011).
42. S. Mukamel, *Annu. Rev. Phys. Chem.* **51**, 691 (2000).
43. D. Abramavicius, B. Palmieri, D. V. Voronine, F. Šanda and S. Mukamel, *Chem. Rev.* **109**, 2350 (2009).
44. Y. J. Yan and S. Mukamel, *J. Chem. Phys.* **89**, 5160 (1988).
45. S. Mukamel, *The Principles of Nonlinear Optical Spectroscopy* (Oxford University Press, New York, 1995).
46. Y. J. Yan and S. Mukamel, *J. Chem. Phys.* **94**, 179 (1991).
47. S. Mukamel and A. Tortschanoff, *Chem. Phys. Lett.* **357**, 327 (2007).
48. S. Mukamel, R. Oszwałdowski and L. Yang, *J. Chem. Phys.* **127**, 221105 (2007).
49. D. Abramavicius, D. V. Voronine and S. Mukamel, *Proc. Natl. Acad. Sci. USA* **105**, 8525 (2008).
50. S. Mukamel, Y. Tanimura and P. Hamm, *Special Issue Acc. Chem. Res.* **42**, 1207 (2009).

CHAPTER 6

FIRST-PRINCIPLES CALCULATIONS FOR LASER INDUCED ELECTRON DYNAMICS IN SOLIDS

K. Yabana*,†, Y. Shinohara†, T. Otobe‡,
Jun-Ichi Iwata§ and George F. Bertsch¶

We calculate the electron dynamics in a crystalline solid in the presence of strong, ultrashort laser pulses within time-dependent density functional theory. The evolution of the electron orbitals is given by the time-dependent Kohn–Sham equation which includes the electric field of the laser pulse. There are several issues that arise in developing a practical computational method to implement the theory. The first is to separate the macroscopic and microscopic length scales in the dynamics. The second is treatment of the medium polarization. Both these are overcome by a suitable choice of the electromagnetic gauge. The resulting computational framework can be applied to a wide variety of optical phenomena in solids. On a purely microscopic scale, the method may be used to calculate dielectric properties of crystalline solids. For an ultrashort laser pulse of moderate intensity, it may be used to investigate mechanisms of coherent phonon generation. As the field strength increases, it may describe dense electron-hole excitations in solids leading to an optical breakdown. At the most general level, the present formalism can treat the fully coupled dynamics between macroscopic electromagnetic fields and microscopic electron dynamics. Although it is extremely demanding computationally, this will provide the most comprehensive description for the interaction of strong and ultrashort laser pulses with solids.

*Center for Computational Sciences, University of Tsukuba, Tsukuba 305-8577, Japan.
†Graduate School of Pure and Applied Sciences, University of Tsukuba, Tsukuba 305-8571, Japan.
‡Advanced Photon Research Center, JAEA, Kizugawa, Kyoto 619-0215, Japan.
§Department of Applied Physics, University of Tokyo, Tokyo 113-8656, Japan.
¶Institute for Nuclear Theory and Department of Physics, University of Washington, Seattle, WA98195, USA.

1. Introduction

When a very short laser pulse of a few tens of femtoseconds or shorter irradiates molecules and solids, the immediate effect of the electric field is to excite the electrons. At this stage of interaction, one may ignore motion of atomic nuclei since vibrational periods are comparable to or even longer than the duration of the laser pulse. For a laser pulse of weak intensity, the electronic response may be treated within linear response theory. The response in the frequency domain can be calculated simply by the Fourier transform of the real-time response. As the laser intensity increases, nonlinear optical responses become significant and high-order perturbation theories would be necessary for a direct calculation in the frequency domain. Under an extremely strong electric field for which the perturbative expansion is no more useful, the only choice is to solve the time-dependent Schrödinger equation as given in the time domain.

In the last two decades, numerical approaches solving time-dependent Schrödinger equation have been developed and utilized extensively in studying electron dynamics induced by strong and ultrashort laser pulses.[1,2,3] A direct calculation of electron dynamics for many-electron systems is, however, only feasible in practice for one or two active electrons. Therefore, one must rely upon an approximate theory for extended systems. Here we are guided by the enormous success of density functional theory (DFT) to describe the structure of matter in their electronic ground states. The time-dependent DFT (TDDFT) provides a reasonable basis for the description of many electrons dynamics.[4-9] The TDDFT is an extension of the DFT so that it may describe either electronic excited states in the linear response framework or nonlinear electron dynamics induced by a strong external field as initial value problems.[10,11] One should be cautioned, however, that there is no systematic way to evaluate the functionals in use except by their performance in reproducing known properties.

In this chapter, we review our theoretical and computational studies to describe electron dynamics in crystalline solids induced by strong and ultrashort laser pulses.[12-16] Since a typical wavelength of the laser pulse is a few hundreds of nm and is much larger than the length of a unit cell in solids, we may assume a long wavelength limit. Namely, we consider electron dynamics in a unit cell under a time-dependent, spatially uniform electric field.[17]

There are two distinct aspects in formulating electron dynamics in a crystalline solid which are absent in isolated systems such as atoms and molecules. One is the choice of the electromagnetic gauge. As we will discuss later, the velocity gauge should be used in describing the electron dynamics in a crystalline solid, since it allows us to apply the Bloch theorem at each time for the electron orbitals.[17] The other is regarding the treatment of the polarization field. In calculating electron dynamics in a crystalline solid, we solve the time-dependent Kohn–Sham (TDKS) equation in a unit cell which is assumed to be located far from the surface. However, the electric field in the unit cell should include the polarization field which is caused by the surface charge and depends on the macroscopic shape of the material considered. Therefore, we need to specify the macroscopic shape of the sample in the calculation.[15]

There are a number of interesting phenomena in light-matter interactions for which our method is useful. Calculating electron dynamics under weak field, our method may be used to calculate linear susceptibilities of crystalline solid such as the dielectric function.[17] Calculating forces acting on atoms induced by ultrashort laser pulses, we may investigate mechanisms of coherent phonon generation, coherent atomic motions in a macroscopic spatial size.[16,18] As the intensity of the laser pulse increases, the high-order nonlinear response becomes significant. When the carrier density of excited electrons reaches a certain critical value, the material finally suffers optical breakdown.[12,13] All of these effects are present in the modeling of the dynamics by TDDFT. The main caveat is that the TDDFT neglects electron–electron collisions, and so it cannot be considered realistic for describing the thermalization of highly excited systems.

The construction of this paper is as follows. In Sec. 2, we explain the theoretical framework. In Sec. 3, we apply the theory to linear response properties. In Sec. 4, we present applications to coherent phonon generation in the semiconductor Si. In Sec. 5, we calculate the energy transfer and the formation of dense electron-hole excitations by an intense laser field in diamond. In Sec. 6, we discuss an extension of our scheme to describe a coupled dynamics of macroscopic electromagnetic fields and microscopic electron dynamics. Finally, Sec. 7 summarizes our work.

2. Formalism

In this section, we outline our TDDFT formalism for describing electron dynamics in a crystalline solid. The next section presents an intuitive description of the TDKS equation that we solve. The equation is derived in the following section using a Lagrangian formalism.

2.1. *A time-dependent Kohn-Sham equation in periodic systems*

We consider a large but finite system under a time-dependent, spatially uniform electric field, $\mathbf{E}(t)$. The TDKS equation for electrons in the length gauge is given by

$$i\hbar \frac{\partial}{\partial t} \psi_i(\mathbf{r}, t) = \left\{ \frac{\mathbf{p}^2}{2m} + V_{\text{ion}}(\mathbf{r}) + \int d\mathbf{r} \frac{e^2}{|\mathbf{r} - \mathbf{r}'|} n(\mathbf{r}') \right.$$

$$\left. + \mu_{xc}[n(\mathbf{r}, t)] + e\mathbf{E}(t) \cdot \mathbf{r} \right\} \psi_i(\mathbf{r}, t), \tag{2.1}$$

where $\psi_i(\mathbf{r}, t)$ is a TDKS orbital and $n(\mathbf{r}, t)$ is the electron density, related to the orbitals by $n(\mathbf{r}, t) = \sum_i |\psi_i(\mathbf{r}, t)|^2$. The $\mu_{xc}[n(\mathbf{r}, t)]$ is the exchange-correlation potential. The interaction of electrons with the electric field is given by $e\mathbf{E}(t) \cdot \mathbf{r}$.

We move to the velocity gauge by making a gauge transformation,

$$\psi_i(\mathbf{r}, t) = \exp\left[\frac{ie}{\hbar c} \mathbf{A}(t) \cdot \mathbf{r} \right] \tilde{\psi}_i(\mathbf{r}, t), \tag{2.2}$$

where the vector potential as a function of time, $\mathbf{A}(t)$, is defined by

$$\mathbf{A}(t) = -c \int^t dt' \mathbf{E}(t'). \tag{2.3}$$

The TDKS equation in the velocity gauge is given by

$$i\hbar \frac{\partial}{\partial t} \tilde{\psi}_i(\mathbf{r}, t) = \left\{ \frac{1}{2m} \left(\mathbf{p} + \frac{e}{c} \mathbf{A}(t) \right)^2 + \tilde{V}_{\text{ion}}(\mathbf{r}) + \int d\mathbf{r} \frac{e^2}{|\mathbf{r} - \mathbf{r}'|} n(\mathbf{r}', t) \right.$$

$$\left. + \mu_{xc}[n(\mathbf{r}, t)] \right\} \tilde{\psi}_i(\mathbf{r}, t). \tag{2.4}$$

Equations (2.1) and (2.4) are equivalent in finite systems, but as we mentioned earlier the velocity gauge equation (2.4) is convenient for periodic systems. To understand it, we consider a very large but finite crystalline solid in which the unit cell of the material is characterized by the three unit-cell vectors, $\mathbf{a}_i$, ($i = 1 - 3$). In a unit cell far apart from the surface of the material, the ionic potential is periodic in space, $\tilde{V}_{\text{ion}}(\mathbf{r} + \mathbf{a}_i) = \tilde{V}_{\text{ion}}(\mathbf{r})$. Under a spatially uniform electric field, we may expect that the time-dependent orbitals remain spatially periodic.

However, in the length gauge, the Hamiltonian in Eq. (2.1) is not periodic due to the linear potential, $e\mathbf{E}(t) \cdot \mathbf{r}$. We also note that even the equation itself is not defined well in the limit of large system since the potential diverges at infinity. In the velocity gauge, on the other hand, the Hamiltonian in Eq. (2.4) can be periodic in space, $h(\mathbf{r} + \mathbf{a}_i) = h(\mathbf{r})$, in the presence of the spatially uniform electric field. Therefore, we may apply the Bloch theorem at each time t. We thus adopt the velocity gauge, Eq. (2.4), as the most convenient starting point for the electron dynamics in periodic systems.

2.2. Polarization field

The next issue is the treatment of the polarization field. We again consider a large but finite system irradiated by a spatially uniform, time-dependent electric field. The applied electric field produces an induced polarization and a surface charge in the material. At the unit cell far apart from the surface, the vector potential $\mathbf{A}(t)$ in the TDKS equation (2.4) is the sum of the two contributions, the externally applied electric field and the induced polarization field. The magnitude of the polarization field depends on the macroscopic shape of the material. Therefore, in solving the TDKS equation (2.4) in a unit cell, we need to specify the contribution of the polarization field from outside.

In the following, we will consider two cases shown in Fig. 1 which we will call the transverse and the longitudinal geometries. We consider a solid which is infinitely periodic in x and y directions, and which is sufficiently thick but finite in z direction. In the transverse geometry shown in panel (a) of Fig. 1, the laser pulse propagates in the z direction while the polarization direction of the electric field in the xy plane. In the panel (b) of Fig. 1

Advances in Multi-Photon Processes and Spectroscopy

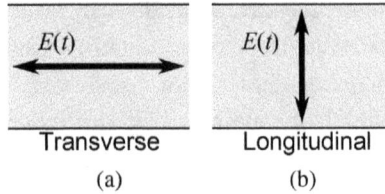

Fig. 1. Electric field applied to a solid which is finite in z direction and infinite in x and y directions. (a) Transverse geometry when the direction of the electric field is perpendicular to z axis, (b) longitudinal geometry when parallel to z axis.

showing the longitudinal geometry, the laser pulse propagates in the xy plane parallel to the surface of the solid and the polarization direction is parallel to z.

In the transverse geometry, there is no surface charge and the polarization field does not contribute to the electric field in the unit cell. In this case, the vector potential $\mathbf{A}(t)$ in Eq. (2.4) is given by the externally applied electric field only,

$$\mathbf{A}(t) = \mathbf{A}_{\text{ext}}(t). \tag{2.5}$$

In the longitudinal geometry, the vector potential $\mathbf{A}(t)$ in Eq. (2.4) is the sum of the externally applied electric field and the induced polarization field,

$$\mathbf{A}(t) = \mathbf{A}_{\text{ext}}(t) + \mathbf{A}_{\text{ind}}(t). \tag{2.6}$$

The polarization field is related to the surface charge $\sigma(t)$ and the macroscopic current, $\mathbf{I}(t)$, flowing inside the bulk solid.

$$\mathbf{E}_{\text{ind}}(t) = -\frac{1}{c}\frac{d\mathbf{A}_{\text{ind}}(t)}{dt} = -4\pi\sigma(t)\hat{z}, \tag{2.7}$$

$$\frac{d\sigma(t)}{dt} = I(t). \tag{2.8}$$

The macroscopic current may be evaluated from the solution of the TDKS equation (2.4) by averaging the microscopic current $\mathbf{j}(\mathbf{r}, t)$ over the unit cell volume,

$$\mathbf{I}(t) = \frac{-e}{\Omega}\int_{\Omega} d\mathbf{r}\,\mathbf{j}(\mathbf{r}, t), \tag{2.9}$$

where Ω is the volume of the unit cell and the microscopic current, $\mathbf{j}(\mathbf{r}, t)$, is given by,

$$\mathbf{j}(\mathbf{r}, t) = \sum_i \frac{1}{2m} \left[\tilde{\psi}_i^*(\mathbf{r}, t) \left(\mathbf{p} + \frac{e}{c}\mathbf{A} \right) \tilde{\psi}_i(\mathbf{r}, t) - c.c. \right]. \qquad (2.10)$$

Equations (2.4) and (2.6)–(2.10) constitute a closed set of equations for the longitudinal geometry.

2.3. *Derivation from a Lagrangian*

In this subsection, we will show that the same equations of motion as those discussed in the previous subsection may be derived in the Lagrangian formalism.[17] In addition to the electron orbitals and electromagnetic fields, we treat the coordinates of ions as dynamical variables. The Lagrangian formalism has a merit that it allows us to construct an expression for the conserved energy of the total system. The energy conservation is quite useful in practical calculations to examine the accuracy of the numerical calculation.

We start with the ordinary Lagrangian for a coupled system of electrons, ions, and electromagnetic fields, except that we employ the TDDFT instead of quantum mechanics for electrons. As basic variables, we use TDKS orbitals, $\psi_i(\mathbf{r}, t)$, scalar and vector potentials, $\phi(\mathbf{r}, t)$ and $\mathbf{A}(\mathbf{r}, t)$, and the coordinate of ions, $R_\alpha(t)$, $(\alpha = 1 \cdots N_I)$. Treating the electromagnetic field and the ionic coordinates classically, the Lagrangian is

$$\begin{aligned}
L = & \sum_i \int d\mathbf{r} \left\{ \psi_i^* i\hbar \frac{\partial}{\partial t} \psi_i - \frac{1}{2m} \left| \left(\mathbf{p} + \frac{e}{c}\mathbf{A} \right) \psi_i \right|^2 \right\} \\
& - \int d\mathbf{r} \, (en_{\text{ion}} - en_e)\phi - E_{xc}[n_e] \\
& + \frac{1}{8\pi} \int d\mathbf{r} \left\{ \left(-\nabla\phi - \frac{1}{c}\frac{\partial \mathbf{A}}{\partial t} \right)^2 - (\nabla \times \mathbf{A})^2 \right\} \\
& + \frac{1}{2} \sum_\alpha M_\alpha \left(\frac{d\mathbf{R}_\alpha}{dt} \right)^2 + \frac{1}{c} \sum_\alpha Z_\alpha e \frac{d\mathbf{R}_\alpha}{dt} \cdot \mathbf{A}, \qquad (2.11)
\end{aligned}$$

where $n_{\text{ion}}(\mathbf{r})$ is the density of ions given by $n_{\text{ion}}(\mathbf{r}) = \sum_\alpha Z_\alpha \delta(\mathbf{r} - \mathbf{R}_a)$. M_α and Z_α are the mass and charge number of ions, respectively. The exchange-correlation functional, $E_{xc}[n]$, has been taken to be the same as in the ground-state functional. This is called the adiabatic approximation. We further assume that $E_{xc}[n]$ is a local function of n, to arrive at the adiabatic local density approximation (ALDA).

We introduce a periodic structure for this Lagrangian as follows. For the ionic coordinates $R_\alpha(t)$ and ionic density $n_{\text{ion}}(\mathbf{r})$, we assume a periodic structure with the three unit-cell vectors, $\mathbf{a}_i$, ($i = 1 - 3$), as before. The scalar potential $\phi(\mathbf{r}, t)$ is assumed to be periodic, while the vector potential is responsible for the spatially uniform part of the electric field. We thus assume that the vector potential is a function of time variable only, $\mathbf{A}(t)$. For the orbital wave functions, we assume that we may apply the Bloch theorem at each time so that we can introduce periodic orbital $u_{n\mathbf{k}}(\mathbf{r}, t)$ by $\psi_i(\mathbf{r}, t) = e^{i\mathbf{k}\mathbf{r}} u_{n\mathbf{k}}(\mathbf{r}, t)$ where the index i is composed of the band index n and the crystal momentum $\mathbf{k}$. Now all the variables in the Lagrangian (2.1) have a periodicity in the unit cell, and we may write it as an integral over unit cell volume Ω.

$$
\begin{aligned}
L = \sum_{n\mathbf{k}} \int_\Omega d\mathbf{r} \left\{ u_{n\mathbf{k}}^* i\hbar \frac{\partial}{\partial t} u_{n\mathbf{k}} - \frac{1}{2m} \left| \left(\mathbf{p} + \mathbf{k} + \frac{e}{c}\mathbf{A} \right) u_{n\mathbf{k}} \right|^2 \right\} \\
- \int_\Omega d\mathbf{r}\{(en_{\text{ion}} - en_e)\phi - E_{xc}[n_e]\} + \frac{1}{8\pi} \int_\Omega d\mathbf{r}(\nabla\phi)^2 \\
+ \frac{\Omega}{8\pi c^2} \left(\frac{d\mathbf{A}}{dt} \right)^2 + \frac{1}{2} \sum_\alpha M_\alpha \left(\frac{d\mathbf{R}_\alpha}{dt} \right)^2 + \frac{1}{c} \sum_\alpha Z_\alpha e \frac{d\mathbf{R}_\alpha}{dt} \cdot \mathbf{A}.
\end{aligned}
$$

(2.12)

Next we write down the Euler–Lagrange equations derived from this Lagrangian. The variation of the orbital functions $u_{n\mathbf{k}}(\mathbf{r}, t)$ gives the TDKS equation,

$$
i\hbar \frac{\partial u_{n\mathbf{k}}}{\partial t} = \frac{1}{2m} \left[\mathbf{p} + \mathbf{k} + \frac{e}{c}\mathbf{A}(t) \right]^2 u_{n\mathbf{k}} - e\phi u_{n\mathbf{k}} + \frac{\delta E_{xc}}{\delta n} u_{n\mathbf{k}}, \qquad (2.13)
$$

where $\mu_{xc} = \delta E_{xc}/\delta n$ is the exchange-correlation potential as in Eq. (2.1). The variation with respect to $\phi(\mathbf{r}, t)$ and $\mathbf{A}(t)$ results in

$$\nabla^2 \phi(\mathbf{r}, t) = -4\pi e[-n_e(\mathbf{r}, t) + n_{\text{ion}}(\mathbf{r}, t)], \tag{2.14}$$

$$\frac{\Omega}{4\pi c^2} \frac{d^2 \mathbf{A}(t)}{dt^2} = \frac{e}{c} \int_\Omega d\mathbf{r}\{\mathbf{j}_{\text{ion}} - \mathbf{j}_e\} - \frac{e^2}{mc^2} N_e \mathbf{A}(t). \tag{2.15}$$

The last equation looks similar to Eqs. (2.7)–(2.10) in the longitudinal geometry. However, Equation (2.15) does not include any externally applied electric field. The applied electric field was taken away in the above derivation when we ignored the spatial dependence of the vector potential. To introduce the externally applied laser field, we divide the vector potential into external and induced ones as in the longitudinal case, $\mathbf{A}(t) = \mathbf{A}_{\text{ext}}(t) + \mathbf{A}_{\text{ind}}(t)$, and replace Eq. (2.15) with

$$\frac{\Omega}{4\pi c^2} \frac{d^2 \mathbf{A}_{\text{ind}}(t)}{dt^2} = \frac{e}{c} \int_\Omega d\mathbf{r}\,\{\mathbf{j}_{\text{ion}} - \mathbf{j}_e\} - \frac{e^2}{mc^2} N_e \mathbf{A}(t). \tag{2.16}$$

The variation with respect to $\mathbf{R}_\alpha(t)$ results in the Newtonian equation for ions,

$$M_\alpha \frac{d^2 \mathbf{R}_\alpha}{dt^2} = -\frac{e}{c} Z_\alpha \frac{d\mathbf{A}}{dt} - \frac{\partial}{\partial \mathbf{R}_\alpha} \int_\Omega d\mathbf{r}\, en_{\text{ion}}\phi. \tag{2.17}$$

The force appearing in the right hand side is a sum of three contributions, the repulsive force among ions, the force coming from the uniform electric field which is described by $\mathbf{A}(t)$, and the force coming from electrons in the instantaneous density distribution, $n_e(\mathbf{r}, t)$.

From the Lagrangian of Eq. (2.12), we may construct the Hamiltonian of the system as

$$H = \sum_{n\mathbf{k}} \int_\Omega d\mathbf{r} \frac{1}{2m} \left| \left(\mathbf{p} + \mathbf{k} + \frac{e}{c}\mathbf{A}\right) u_{n\mathbf{k}} \right|^2$$

$$+ \int_\Omega d\mathbf{r} \left\{ \frac{1}{2}(en_{\text{ion}} - en_e)\phi + E_{xc}[n_e] \right\} + \frac{\Omega}{8\pi c^2} \left(\frac{d\mathbf{A}}{dt}\right)^2$$

$$+ \frac{1}{2} \sum_\alpha M_\alpha \left(\frac{d\mathbf{R}_\alpha}{dt}\right)^2 + \frac{1}{c} \sum_\alpha Z_\alpha e \frac{d\mathbf{R}_\alpha}{dt} \cdot \mathbf{A}. \tag{2.18}$$

In the absence of any externally applied electric field, this expression conserves energy.

2.4. *Computational method*

We solve the TDKS equation in real-time and real-space. To represent the Kohn–Sham orbitals, we employ a uniform Cartesian grid. We also sample k-points of the Bloch states with a uniform grid. For the derivatives of the orbital wave functions, we use a high-order finite difference formula.[19] We use typically nine-points formula for the first and second derivatives. In the calculation, valence electrons are treated explicitly while the core electron effects are included through a norm-conserving pseudopotential. We employ the pseudopotential constructed by a procedure of Troullier and Martins.[21] The nonlocal part of the pseudopotential is treated with the prescription by Kleinman and Bylander. [21]

We should note that the nonlocal potential is transformed by the gauge field as

$$\tilde{V}_{ion}\tilde{\psi}_i(\mathbf{r}, t) = \int d\mathbf{r}' \, e^{-\frac{ie}{\hbar c}\mathbf{A}(t)\cdot\mathbf{r}} V_{ion}(\mathbf{r}, \mathbf{r}') e^{\frac{ie}{\hbar c}\mathbf{A}(t)\cdot\mathbf{r}'} \tilde{\psi}_i(\mathbf{r}', t). \quad (2.19)$$

It is essential to include this gauge dependence to conserve energy in the TDKS equation. The nonlocal potential also contributes to the current.

The time evolution of the Kohn–Sham orbitals is computed using the Taylor expansion method,[22]

$$\psi_i(t + \Delta t) \approx \exp\left[-\frac{i}{\hbar}h_{KS}(t)\Delta t\right] \psi_i(t)$$

$$\approx \sum_{k=0}^{N} \left(-\frac{i}{\hbar}h_{KS}(t)\Delta t\right)^k \psi_i(t). \quad (2.20)$$

If we keep terms more than third-order in the expansion, we may find a certain upper limit for the time step Δt so that the time evolution may be achieved stably if one use the time-step smaller than Δt. In practice, we employ fourth-order expansion in our calculations.

In the following sections, we will show several calculations for bulk silicon. For each Si atom in the unit cell, the four valence electrons are treated dynamically. The geometry of the unit cell is taken to be a simple cubic cell containing eight Si atoms, with a lattice constant of 10.26 au. The unit cell volume is divided into 16^3 spatial lattice to represent the Kohn–Sham orbitals. The k-space is divided into $24^3 k$-space grid unless otherwise specified. The time-step is chosen as $\Delta t = 0.08$ au.

3. Real-Time Calculation for Dielectric Function

The TDDFT has been most successful in describing electronic excitations and optical properties of atoms and molecules. In these applications, the electronic response of a system to a small perturbation is described by TDDFT in the linear response theory. In the linear response calculations, one does not need to solve the TDKS equation in time domain. One usually recasts the problem into an eigenvalue equation for the electronically excited states[23] or into a linear algebraic equation for the response to an external perturbation of a given frequency.[24,25] Nevertheless, the time-domain calculation of the TDKS equation is an efficient alternative for the linear response calculations, in particular when one is interested in the photoionization process.[22,26,27]

The linear response TDDFT seems less successful for optical responses of solids than those for molecules. This is in part related to the well-known defect of the local density approximation (LDA) that band-gap energies of dielectrics are systematically underestimated. There have been extensive efforts to find a better energy functional to overcome this problem. Indeed, the problem is resolved to some extent if one employs improved functionals including nonlocal exchange effects.[28] In this chapter, we mostly show results with the simple ALDA functional, focusing computational aspects of describing electron dynamics in time domain.

We will use both transverse and longitudinal geometries in the time-domain linear response calculations. Although time evolutions of orbitals are different between two geometries, calculated physical quantities in the linear response approximation accurately coincide, as they should.

3.1. *Linear response calculation in transverse geometry*

We explain how we may calculate frequency dependent dielectric function from a real-time calculation of electron dynamics in the transverse geometry.[15] From the calculated time evolution, we obtain the macroscopic current induced by the external field, $J(t)$. This is a spatial average of the microscopic current over the unit cell volume (see Eq. (2.9)). The induced current is related to the external vector potential through a conductivity as a function of time, $\sigma(t)$, as

$$J(t) = \int^t dt' \sigma(t - t') E(t') = -\frac{1}{c} \int^t dt' \sigma(t - t') \frac{dA(t')}{dt}. \qquad (3.1)$$

Taking the Fourier transformation, the conductivity as a function of frequency may be calculated from the Fourier transformations of the induced current and the applied electric field,

$$\sigma(\omega) = \frac{\int dt\, e^{i\omega t} J(t)}{\int dt\, e^{i\omega t} E(t)}. \qquad (3.2)$$

Since this relation is fulfilled for any weak electric field, one may use an arbitrary electric field, $E(t)$, to determine the conductivity function from the real-time results. The calculated conductivity $\sigma(\omega)$ should not depend on the choice of the electric field, if the calculation is successful. A convenient choice in practice is the impulsive field,

$$A(t) = A_0 \theta(t). \qquad (3.3)$$

For this choice, the conductivity is proportional to the current, $\sigma(t) = -cJ(t)/A_0$. The dielectric function may be obtained from the conductivity from the familiar formula,

$$\varepsilon(\omega) = 1 + \frac{4\pi i \sigma(\omega)}{\omega}. \qquad (3.4)$$

3.2. Linear response calculation in longitudinal geometry

We next consider the same problem in the longitudinal geometry.[17] In this geometry, the vector potential in the TDKS equation is composed of the external and the induced potentials, $A(t) = A_{ext}(t) + A_{ind}(t)$. The total field $A(t)$ is related to the external field $A_{ext}(t)$ through the inverse dielectric function as,

$$A(t) = \int^t dt' \varepsilon^{-1}(t - t') A_{ext}(t'). \qquad (3.5)$$

Taking the Fourier transformation of both sides, we obtain an expression for the frequency-dependent dielectric function directly in terms of the Fourier transformations of external and total vector potentials,

$$\frac{1}{\varepsilon(\omega)} = \frac{\int dt\, e^{i\omega t} A(t)}{\int dt\, e^{i\omega t} A_{\text{ext}}(t)}. \tag{3.6}$$

This relation is again fulfilled for any external vector potentials if the external field is weak enough. The impulsive electric field given by Eq. (3.3) is a convenient choice also in the longitudinal geometry for practical calculations.

3.3. *Example: Dielectric function of bulk Si*

As an example of a calculation of linear response in time domain, we take bulk Si. Figure 2 shows the time profile of the current as a function of time when an impulsive field of Eq. (3.3) is applied at $t = 0$. The red solid curve is the current in the transverse geometry, while the green dashed curve is the current in the longitudinal geometry. Immediately after the impulsive electric field is applied to the solid, all electrons are accelerated coherently. As seen from the figure, the current in the longitudinal geometry shows oscillations of much higher frequency than the current in the transverse geometry. The difference originates from the longitudinal plasma oscillation. In the longitudinal geometry, the current

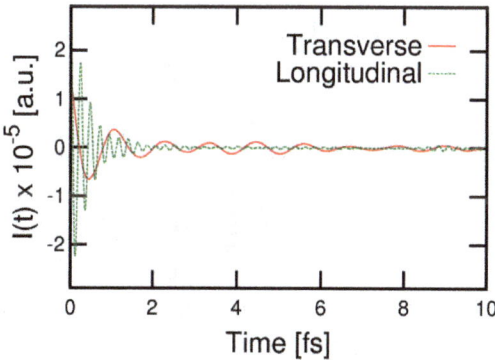

Fig. 2. (Color online) Current averaged over a unit cell volume after the impulsive electric field is applied to crystalline Si. Red curve for the transverse geometry and green curve for the longitudinal geometry.

flowing in the unit cell accumulates at the surfaces to produce the restoring force that causes the longitudinal plasma oscillation. One may verify that the frequency of the oscillation coincides with that of the simple plasma formula, $\omega_p^2 = 4\pi e^2 n/m$, taking the density n from the number of average active electrons in the unit cell, $n = 32/\Omega = 200\,\mathrm{nm}^{-3}$. The current in the transverse geometry shows a much slow oscillation with a period of about 50 au. This period corresponds approximately to the peak of the conductivity just above the direct band-gap of 2.4 eV, to be seen in Fig. 3.

Taking Fourier transformations of the currents, we obtain the dielectric function. Figure 3 shows the result obtained from the calculation in the transverse geometry. The frequency-dependent conductivity calculated by Eq. (3.2) is shown in the left panels, and the dielectric function obtained by Eq. (3.4) is shown in the right panels. The dielectric function obtained from the calculation in the longitudinal geometry Eq. (3.6) coincides accurately with that obtained from the transverse geometry calculation. In the panels of dielectric function, the measured dielectric function is shown as well. The static value of the dielectric function is 12.6 in the ALDA, which is close to the measured value of 11.6.

There are several unsatisfactory features in the calculated dielectric function with ALDA. Among them, the too small band-gap is evident. The measured direct band-gap is 3.3 eV, while the calculated value is 2.4 eV.

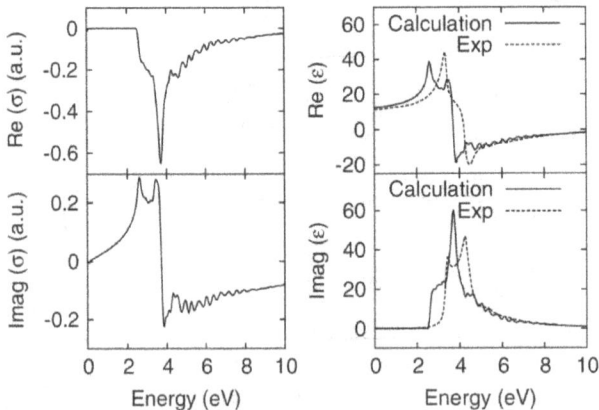

Fig. 3.   Conductivity and dielectric function of Si as functions of frequency calculated by the real-time method. Measured values are also shown for the dielectric function. Taken from Ref. 15.

The double peak structure in the measured Im $\varepsilon(\omega)$ is not reproduced in the calculation. It should be mentioned that there are improved theoretical methods that give a better description of the dielectric function. In particular, solving the Bethe–Salpeter equation with the GW approximation for quasi-particle states is a very successful approach for dielectric functions.[29] Within the TDDFT framework, it has been reported that the hybrid functional incorporating nonlocal exchange potential improves the result.[28] Also, calculations have been recently reported for GW theory using the real-time approach.[30]

4. Coherent Phonon Generation

4.1. *Physical description*

Among phenomena in the interaction of high-intensity ultrashort laser pulses with crystalline solids, coherent optical phonons have been extensively measured in the pump-probe experiments for various solids.[31,32] Coherent phonons can be observed when the duration of the exciting laser pulse is shorter than the period of theoretical phonon. The generation of a coherent phonon starts with the excitation of electrons by the pump pulse. The excited electrons induce a force on atoms through the electron–phonon interaction. The atomic motion is coherent both temporarily and spatially on the macroscopic scale. Observation of the phonon relies on the change of the refractive index of the solid due to the atomic motion. The probe pulse measures the change of the index of refraction and shows an oscillatory pattern whose frequency coincides with the known optical phonon frequency of the crystalline solid.

For a given phonon mode, the reflectivity change is often parameterized by the functional form,

$$\frac{\Delta R}{R} = g e^{-\Gamma t} \cos(\omega t + \phi), \tag{4.1}$$

where g is the amplitude, ω is the phonon angular frequency, Γ is a damping constant, and ϕ is a phase angle. The phase angle ϕ is very sensitive to the mechanism of the coherent phonon generation.

In many theoretical investigations, mechanisms of coherent phonon generation are discussed starting with the following Newtonian equation for a phonon coordinate q,[32]

$$\frac{d^2q}{dt^2} + \omega_{ph}^2 q = F(t), \qquad (4.2)$$

where $F(t)$ is the force acting on ions evaluated at the equilibrium positions of atoms in the ground state and ω_{ph} is the frequency of the optical phonon. There are two distinct mechanisms which have been considered for the coherent phonon generation. They are classified according to whether the electronic excitation caused by the pump pulse is *virtual* or *real*.

In virtual electronic excitations, the electronic state in the solid is distorted during the irradiation of the laser pulse and returns to the ground state after the laser pulse ends. The force $F(t)$ appears only during the irradiation of the laser pulse, and vanishes after the laser pulse ends. The phase ϕ is equal to $\pi/2$ in this mechanism. This mechanism is often called the impulsively stimulated Raman scattering (ISRS).[33]

In real electronic excitations, the orbitals in the unit cell remain excited even after the pump pulse ends. Then the equilibrium positions of atoms change from those in the ground state, and the force $F(t)$ persists after the pump pulse ends. The phase ϕ is equal to zero in this case. This mechanism is often called the resonant Raman scattering.[34] There is also a mechanism accompanying the real electronic excitations called the displacive excitation of coherent phonon (DECP).[35] In this mechanism, an equilibration among electronically excited states is further assumed.

For the ISRS mechanism, one may derive a simple expression for the force under adiabatic approximation. We consider a dielectric of nonzero direct band-gap and assume that the frequency of the laser pulse is small compared to the band-gap. Then we can show that the energy of the solid is lowered by the amount, $E_2 = \chi |E|^2 / 2$, where χ is the dielectric susceptibility tensor. Differentiating this energy with the phonon coordinate, we arrive at the formula for the force,

$$F(t) = \frac{1}{2} \frac{\partial \chi}{\partial q} |E(t)|^2 . \qquad (4.3)$$

We may show that the force in TDDFT, the right-hand side of Eq. (2.17), coincides with Eq. (4.3) in the adiabatic limit.[14]

Beyond the adiabatic regime, Merlin and co-workers proposed an approximate formula called the two-tensor model regarding the resonant Raman scattering mechanism.[34]

4.2. *TDDFT calculation for Si*

We have shown that both mechanisms of virtual and real electronic excitations for the coherent phonon generation are included in the framework of the TDDFT. This may be demonstrated taking again bulk Si as an example.[14] We first illustrate the real-time electron dynamics induced by the ultrashort laser pulse, and then examine the force acting on ions in the TDDFT. Calculations shown below are carried out in the longitudinal geometry. In Fig. 4, we show time profiles of the electric fields for several pulses with different frequencies. The red curve is the externally applied electric field, $E_{ext}(t)$. The laser pulse is taken to have the form,

$$E_{ext}(t) = E_0 \sin^2\left(\frac{\pi t}{T}\right) \sin \omega t, \qquad (4.4)$$

where E_0 is chosen so that the peak intensity of the electric field is equal to $I = 1.0 \times 10^{12}$ W/cm^2. The pulse duration is $T = 16$ fs. The panels (a)–(d) of Fig. 4 correspond to the laser frequencies of $\hbar\omega = 1.0$ eV, 2.5 eV, 3.5 eV, and 6.0 eV, respectively. The green dashed curve is the total electric field inside the solid multiplied with a number indicated in the figure. In the longitudinal geometry, the total electric field is the sum of the external and the induced electric fields, $E(t) = E_{ext}(t) + E_{ind}(t)$. The external and the total electric fields are related by the dielectric function. In panels (a), the laser frequency is much below the direct band-gap energy. We find the total electric field multiplied with 14.2 almost coincides with the external field. This multiplicative factor coincides well with the value of the dielectric function at the laser frequency of 1.0 eV, as seen from Fig. 3. At the laser frequency of 2.5 eV shown in the panel (b), we still find that the external and the total fields are proportional. In (c), we find difference in phase between the external and the total electric fields, indicating complex value of the dielectric function. In (d), the electric fields are again in phase but the multiplicative factor is negative. The observed behavior of the multiplicative factors is consistent with the behavior of the frequency-dependent dielectric function in the TDDFT which was shown in Fig. 3.

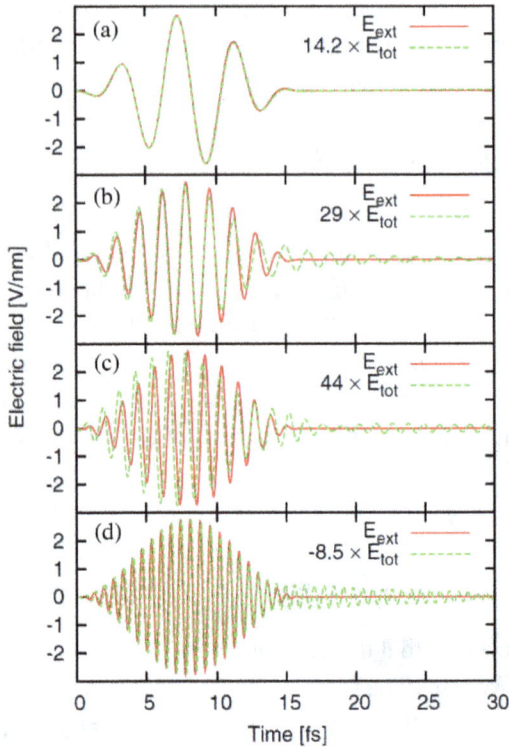

Fig. 4. (Color online) Dielectric response of Si. The red solid line shows the electric field associated with an externally applied laser pulse of the form of Eq. (4.4). The green dashed line shows the total electric field inside the crystal, scaled up by a factor to facilitate the comparison. The panels show results for different laser frequencies. The frequencies (scaling factors) are as follows: (a) 1.0 eV (14), (b) 2.5 eV (29), (c) 3.5 eV (44), (d) 6.0 eV (−8.5). Taken from Ref. 14.

We next show in Fig. 5 the electron density in real-space. The left panel shows the electron density in the ground state of Si in the [110] plane. The right two panels show the change of electron density from that in the ground state. The calculation is shown for the laser pulse of panel (b) of Fig. 4, the laser frequency of 2.5 eV which roughly coincides with the energy of direct band-gap. The middle and right panels correspond to the time, $t = 8.1$ fs and $t = 26.7$ fs, respectively. In the middle and right panels, the red and blue regions indicate an increase or decrease in the electron density, respectively. At $t = 8.1$ fs, the electric field is

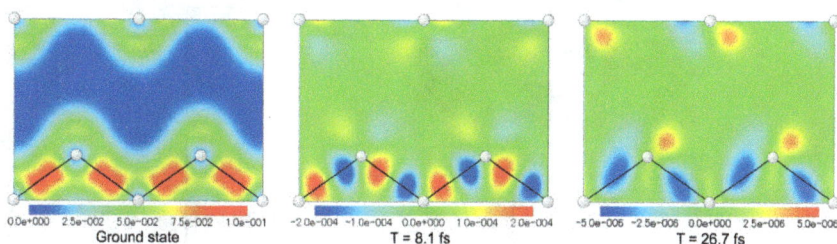

0.0e+000 2.5e-002 5.0e-002 7.5e-002 1.0e-001   -2.0e-004 -1.0e-004 0.0e+000 1.0e-004 2.0e-004   -5.0e-006 -2.5e-006 0.0e+000 2.5e-006 5.0e-006
Ground state   T = 8.1 fs   T = 26.7 fs

Fig. 5. (Color online) Left panel shows the ground-state electron density of Si in [110] plane. The middle and right panels show the change in the electron density from that in the ground state by the laser pulse corresponding to the panel (b) of Fig. 4. The middle panel corresponds to the time $t = 8.1$ fs and the right panel to the time $t = 26.7$ fs, respectively. In the middle and right panels, the red color indicates the increase in the electron density while blue color indicates the decrease. Note that the coloring of the middle and right panels are different by a factor of 40 to improve visibility of the density change at $t = 26.7$ fs. Taken from Ref. 16.

maximum. In the middle panel, a movement of electrons is seen in the bond connecting two Si atoms. At $t = 26.7$ fs, the external electric field already ended. Since the frequency of the electric field, 2.5 eV, is slightly higher than the direct gap energy, 2.4 eV, the external electric field induces real electronic excitations. In the right panel, a decrease of electron density is seen in the bond region connecting two Si atoms. The density change at $t = 26.7$ fs is substantially smaller than that at $t = 8.1$ fs. The density change at $t = 8.1$ fs, which corresponds to the maximum electric field, is caused by the virtual electronic excitations, while the density change at $t = 26.7$ fs, after the externally applied electric field ends, is caused by the real electronic excitations. It is clear from Fig. 5 that the spatial density change is qualitatively different between the virtual and the real excitations.

We next show the force acting on phonon coordinate in Fig. 6. The force is shown for the externally applied electric fields of three frequencies across the direct band-gap energy of 2.4 eV: red curve for 2.25 eV, green curve for 2.5 eV, and blue curve for 2.75 eV. At the lowest frequency shown by the red solid curve, the envelope of the force follows the square of the electric field. Since the laser frequency is below the direct band-gap, electronic excitation is virtual so that the force appears only during the irradiation of the laser pulse. The ISRS mechanism applies in this case. For the laser frequency above the direct band-gap energy, the force keeps a constant value even after the laser pulse ends. Since the laser frequency is above the direct band-gap

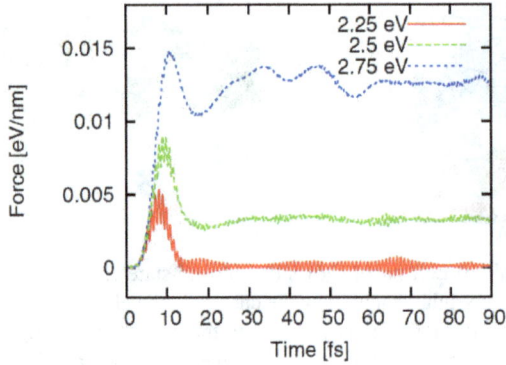

Fig. 6.   (Color online) The force on the optical phonon coordinate for three laser intensities, 2.25 eV (red solid), 2.5 eV (green dashed), and 2.75 eV (blue dotted). Taken from Ref. 14.

energy, real electronic excitation is possible and dominates in this case. For the laser frequency of 2.5 eV, both virtual and real excitations contribute to the force. We thus find that the TDDFT describes both virtual and real electronic excitations that may be responsible for generating coherent phonons.

We next solve the Newtonian equation for the phonon coordinate, Eq. (4.2), with the force shown in Fig. 6. We then fit the calculated time profile $q(t)$ with the formula,

$$q(t) = -q_0 \cos(\omega_{ph} t + \phi) + \bar{q}, \tag{4.5}$$

where the phonon amplitude q_0, the phase ϕ, and the shift of the equilibrium position $\bar{q}$ are determined. The phonon frequency ω_{ph} is set to the measured value, 15.3 THz. The sign of the q_0 is set to $q_0 > 0$. Calculated values of these parameters are summarized in Fig. 7 together with the measured values for the phase.[36–38] For frequencies below the direct band-gap 2.4 eV, the amplitude remains almost constant, the phase is $\pi/2$, and the shift is almost zero. These are consistent with the ISRS mechanism. For frequencies above the direct band-gap, the real electronic excitations become important and dominate the coherent phonon generation. The phase abruptly changes from $\pi/2$ to zero, and the equilibrium position also shifts from that in the ground state. Around the laser frequency of 4.75 eV, we find a vanishingly small amplitude of the phonon. At this frequency, we also find a change of sign of the equilibrium position. As seen from the right panel of Fig. 5,

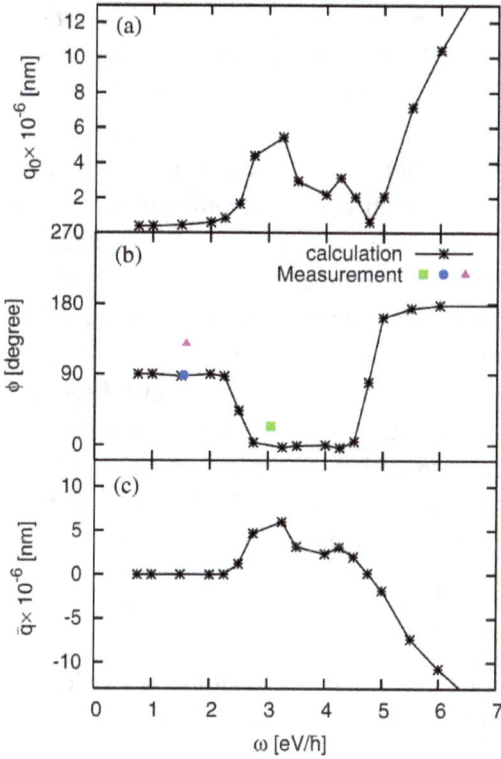

Fig. 7. (a) The amplitude q_0, (b) the phase ϕ, and (c) the shift $\bar{q}$ of the phonon oscillation of Eq. (4.5) as a function of laser frequency. [Taken from Ref. 16].

the real electronic excitation accompanies decrease of the electrons from the bond region. This removal of bonding electrons induces elongation of the bond in the plane where the polarization direction of the laser electric field is included. However, for laser frequencies around 4.75 eV, a number of electron-hole pairs contribute to the force and cancel their contributions. Above that frequency, the net contribution of the electron-hole pairs gives the force opposite in the direction to the lower frequency case.

Summarizing calculations for the coherent phonon generation in Si, the TDDFT is capable of describing two distinct mechanisms caused by virtual and real electronic excitations. For laser frequencies below the direct band-gap, the electronic structure is distorted virtually by the presence of the electric field. The distortion induces the force acting on optical phonon. Since the electronic state returns to the ground state after the laser pulse ends, the force appears only during the irradiation of the laser pulse. This corresponds to the ISRS mechanism. For laser frequencies above the direct band-gap, real electronic excitations take place by the photo-absorption. The excitations continue even after the laser pulse ends. In real-space, a decrease of bonding electrons is seen. It induces bond weakening, and causes the force acting on phonon coordinate which persists even after the laser pulse ends. The virtual electronic excitation also contributes to the force for the laser frequencies above the band-gap. However, the real excitation mechanism dominates since the force caused by the real excitations is much stronger than that by the virtual excitations.

Measurements of coherent phonons have been made for various materials besides dielectrics like Si. The coupling to the optical phonon mode is especially strong in the semimetals, Bi and Sb, see e.g. Ref. 39. We have recently carried out TDDFT calculation in Sb, where the band structure permits the real electronic excitations even at low frequency.[16]

5. Optical Breakdown

As the intensity of the laser pulse irradiating on a dielectric increases, more electron-hole pairs are created in the dielectric. When the density of excited electrons exceeds a certain threshold, the dielectric suffers an irreversible damage. This phenomenon is called optical breakdown.

For a laser pulse much longer than tens of femtosecond, the avalanche mechanism induced by electron–electron collisions is supposed to be mainly responsible for the production of dense electron-hole pairs.[40] As the duration of the laser pulse decreases, direct excitations of electrons either by multiphoton or tunneling ionizations become more and more important.[41] In the TDDFT with ALDA, only the latter processes of multiphoton and tunneling ionization are accessible by the theory. Therefore, the present TDDFT with ALDA approach is expected to be adequate to describe electron dynamics in an extremely short period where electron–electron collisions are not yet important.

5.1. *Incident, external, and internal electric fields*

As we mentioned in Sec. 2.2, the electric field inside a solid depends on the macroscopic shape of the sample and the propagation direction of the laser pulse. In experiments, the intensity of the laser pulse is usually characterized by the maximum value of the incident electric field. In our TDDFT calculations, we may specify the maximum value of the applied electric field in either the longitudinal or the transverse geometries. We first consider relationship between the maximum electric fields of the applied laser pulses in experiments and in our calculations. We assume that electric fields are sufficiently weak that we may apply the ordinary boundary condition in electromagnetism for the electric fields at the surface of the solid.

Let us first consider the case of the transverse geometry in which a linearly polarized laser pulse irradiates the bulk solid at normal incidence. Since the polarization direction is parallel to the surface, the electric field is continuous across the surface. To establish the relation between electric fields of inside and outside the solid, we need to take into account the reflection of the incident pulse at the surface. The relation between the incident electric field from outside the medium, E_{in}, and the electric field in the medium, E_{medium}, is given by

$$E_{medium} = \frac{2}{1 + \sqrt{\varepsilon}} E_{in}, \tag{5.1}$$

where ε is the dielectric constant of the medium.

We next consider the case of the longitudinal geometry in which a linearly polarized laser pulse propagates in parallel to the surface of the solid. The polarization direction of the electric field is perpendicular to the surface. Then the electric fields inside and outside the solid is not continuous because of the surface charge. The amplitude of the electric field outside the medium which is equal to that we called the externally applied electric field, E_{ext}, in Eq. (2.6), is related to the electric field inside the medium, E_{medium}, by

$$E_{ext} = \varepsilon E_{medium}. \tag{5.2}$$

When we discuss the intensity of the laser pulse in making a comparison between theoretical results and measurements, we need to be careful how the magnitude of the electric field is specified in the calculation. In our early publications, we adopted the longitudinal geometry in our calculation and indicated the intensity of the applied laser pulse, E_{ext}, of Eq. (2.6).[12,13]

As is evident in the above argument, the intensity of the applied electric field used in the calculation in longitudinal geometry does not correspond correctly to the intensity of the incident laser pulse in experiments.

We also note that the above argument relating the experimental laser intensity and the strength of the electric field inside solids is valid for weak laser pulses, since all quantities are determined by the dielectric function. For intense laser pulses which may create a number of electron-hole pairs at the solid surface, the dielectric function changes from that in the ground state so that Eqs. (5.1) and (5.2) no longer apply. An extension of the theoretical framework to be discussed in the next section provides a scheme where we may relate intensities of the laser pulse in experiments and in the calculation in a natural way.

5.2. *Intense laser pulse on diamond*

In Otobe *et al.* and Otobe *et al.*,[12,13] we presented calculations of electrondynamics under intense field in the longitudinal geometry for diamond and α-quartz, respectively. In Yabana *et al.*,[15] we reported a comparison of longitudinal and transverse calculations for Si, adding to a multiscale calculation which will be explained briefly in the next section.

In Fig. 8, we show calculated results for diamond. The laser frequency is set to 3.1 eV. In the present ALDA calculation, the band-gap is 4.8 eV, which is smaller than measured value of about 7 eV. Blue curves show externally applied electric field, $E_{ext}(t)$, and red curves show the sum of the applied electric field and the polarization, $E_{tot}(t)$, for several intensities of the external electric fields. We note that the intensities indicated in the figure are those evaluated from the maximum value of the external electric field and do not correspond to the laser intensity in experiments. For weak laser pulses, the two fields are related by the inverse dielectric function,

$$E_{tot}(t) = \int^{t} dt' \varepsilon^{-1}(t - t') E_{ext}(t'). \quad (5.3)$$

In the figure, we find two fields are mostly proportional for intensities up to 5×10^{14} W/cm^2. In the longitudinal geometry adopted in the present calculation, the electric field inside the solid is screened by the surface charge. The total electric field is smaller than the external one by about a factor of 5, which is close to the value of the dielectric function at 3.1 eV

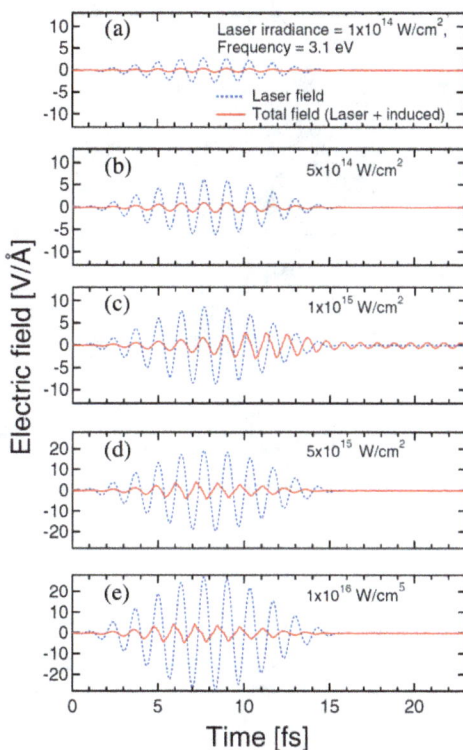

Fig. 8. (Color online) Electric fields of the externally applied laser pulse (blue dotted curve) and the total electric fields (red solid curve) are shown as a function of time for different laser intensities. Calculation is done for diamond at the frequency of 3.1 eV. Taken from Ref. 12.

in the TDDFT calculation. At 1×10^{15} W/cm^2, the response is linear at the beginning. However, as the intensity of the electric field increases, the screening effect becomes weakened. We also find a shift in phase between the external and the total fields. This indicates the energy transfer from the external electric field to electrons in the solid. Above 1×10^{15} W/cm^2, the phase between the external electric field and the total field becomes $\pi/2$ and more. This indicates that a strong absorption of the laser pulse takes place.

We next examine electron dynamics in some detail when the phase difference between the external and total electric fields appears. In Fig. 9, we show electric fields (top), the number density of excited electrons (middle),

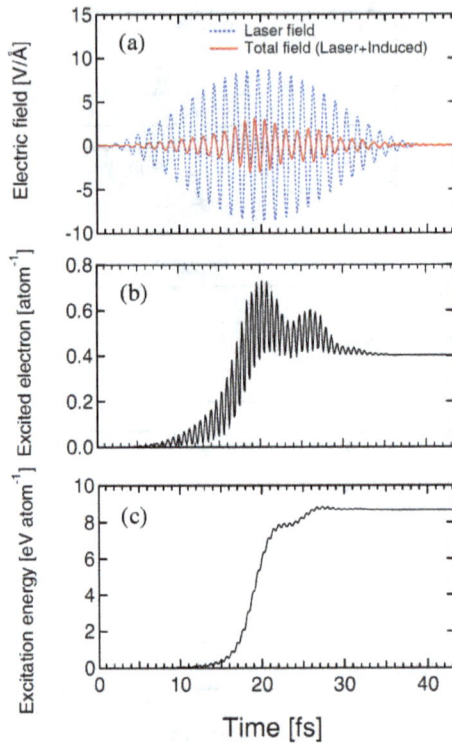

Fig. 9. (Color online) Electric field of an externally applied laser pulse (blue dashed curve) and the total electric field (red solid curve) are shown in (a) as a function of time. The external laser pulse is characterized by a maximum intensity of 1×10^{15} W/cm^2, a pulse duration of 40 fs, and a laser frequency of 3.1 eV. The number of photoexcited electrons per carbon atom and the excitation energy per carbon atom are shown in (b) and (c), respectively. Taken from Ref. 12.

and electronic excitation energy per unit volume (bottom). The maximum intensity of the external laser pulse is set to 1×10^{15} W/cm^2. At the initial stage of the laser pulse irradiation, the external and the total electric fields are in phase. Starting at 15 fs, the number of excited electrons and the excitation energy per unit volume undergo a rapid increase. Simultaneously, the total electric field starts to go out of phase with the external electric field, signaling a large energy transfer. By about 20 fs, the external and the total electric fields are completely out of phase. At this point, the number of excited electrons and the excitation energy reach their saturation value.

These behaviors of electrons and polarization field may be understood in the following way. In the first stage, the number of excited electrons increases as the magnitude of the electric field increases. Since there is no band-gap for electrons excited into conduction bands, the excited electrons may behave as metallic and show a collective plasma oscillation. One may estimate the plasma frequency by

$$\omega_p = \left(\frac{4\pi n_{ex}}{m^* \varepsilon} \right)^{1/2}, \tag{5.4}$$

where ε is the dielectric constant of the diamond and m^* is the electron-hole reduced mass. As seen from Fig. 9(b), the saturated value of the excited electron density is given by 0.4/atom. If we assume a free electron mass for m^*, the corresponding plasma frequency is $\hbar\omega_p = 4.0\,\text{eV}$, which is slightly higher than the frequency of the externally applied electric field ($\hbar\omega_p = 3.1\,\text{eV}$).

This resonant condition explains the behavior shown in Fig. 9. As the intensity of the external electric field increases, electrons excited into conduction bands increase. When the density reaches a critical value so that the plasma frequency coincides with the frequency of the external laser pulse, a resonant energy transfer occurs from the laser pulse to electrons.

In Fig. 10, we show the energy transfer from laser pulse to electrons in diamond as a function of laser intensity. Figure 10(a) is taken from Fig. 4 of Ref. 12. The laser frequency is again fixed at 3.1 eV. The laser intensity in this figure is evaluated from the externally applied electric field, $E_{ext}(t)$, of Eq. (5.2). As discussed in Sec. 5.1, this intensity does not correspond to the intensity of the incident laser pulse in experiments. Assuming a normal incidence of the laser pulse, the intensity of the incident laser pulse, I_{in}, and that of the externally applied electric field in the longitudinal geometry, I_{ext}, are related by,

$$I_{ext} = \left(\frac{2\varepsilon}{1 + \sqrt{\varepsilon}} \right)^2 I_{in}. \tag{5.5}$$

Putting static value of dielectric constant, $\varepsilon \approx 6$, the factor in Eq. (5.5) is about 12.1. Figure 10(b) is the same as Fig. 10(a) except that the laser intensity in the horizontal axis is changed by this factor. To make a comparison with measurements, Fig. 10(b) should be used. Since the linear

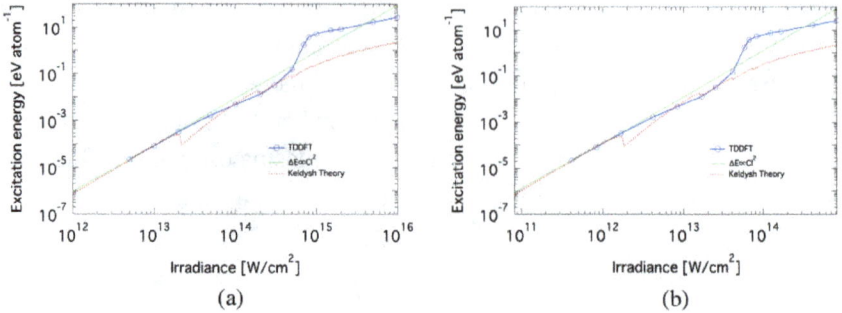

Fig. 10. (Color online) The energy deposited in a diamond by a laser pulse is shown as a function of intensity at a fixed laser frequency of 3.1 eV. The calculated values are shown by open circles connected with a blue line. The laser intensity of an applied electric field, $E_{ext}(t)$, is used in the left panel (a), while the corresponding intensity of the incident laser pulse by Eq. (5.5) is used in the right panel (b). A curve of $\Delta E \propto I^2$ dependence is shown by a green dotted line. An estimation by the Keldysh theory is also plotted by a red dashed line. The curves of a quadratic dependence and of the Keldysh theory are normalized so that they coincide with the value of a real-time calculation at 5×10^{12} W/cm^2 in (a). Left panel (a) from Ref. 12.

relation is assumed to derive Eq. (5.5), the horizontal axis of Fig. 10(b) may not be reliable when nonlinearity in electronic responses becomes significant.

Since at least two photons are required for the valence electrons to be excited into conduction bands, one may expect that the energy transfer, ΔE, depends on the laser intensity I as $\Delta E \propto I^2$. The corresponding curve is drawn in Figs. 10(a) and 10(b) by green dotted line. A theoretical estimation based on the Keldysh theory is also drawn by the red dashed curve. The curves of quadratic dependence and of the Keldysh theory are normalized to the real-time calculation at the intensity of 5×10^{12} W/cm^2 in Fig. 10(a). For a weak intensity region, a calculated energy transfer accurately follows the quadratic dependence. At intensities higher than 6×10^{13} W/cm^2 in Fig. 10(b) (7×10^{14} W/cm^2 in Fig. 10(a)), the energy transfer shows an abrupt increase. This behavior is consistent with the occurrence of a resonant energy transfer discussed above.

We may tentatively assign the abrupt increase of the energy transfer to the optical dielectric breakdown threshold observed in measurements. In Otobe *et al.*,[12] we reported the calculated threshold for the dielectric breakdown to be $6 \, \text{J/cm}^2$, from the critical intensity of $7 \times 10^{14} \, \text{W/cm}^2$ in Fig. 10(a) and the pulse duration of 16 fs. The threshold for the damage in a diamond has been measured at $0.63 \pm 0.15 \, \text{J/cm}^2$ for 2 eV and a 90 fs pulse,[42] which is much lower than the calculation if we make a comparison with measurement using the intensity of the external electric field. However, as is evident from the above argument, we should make a comparison with the laser intensity shown in Fig. 10(b). Then the critical intensity is given by $6 \times 10^{13} \, \text{W/cm}^2$ and the fluence by $0.5 \, \text{J/cm}^2$ in complete agreement with measurement.

We thus conclude that it is quite important to relate carefully the magnitude of the electric field inside a solid with the magnitude of the electric field of incident laser pulse. To do so, we need to take into account the effect of the polarization adequately as well as the reflection of the laser pulse at the surface.

6. Coupled Dynamics of Electrons and Electromagnetic Fields

In the previous sections, we discussed how to describe electron dynamics in a crystalline solid under an electric field whose time profile is prepared in advance. As the intensity of the electric field increases, electron-hole pairs are created at high density in the solid, transferring the energy of the laser pulse to electrons even if the frequency of the laser pulse is below the band-gap. In the extremely strong limit, the propagation of the laser pulse is strongly influenced by the electronic excitations. In such cases, we must solve the coupled dynamics of electrons and electromagnetic fields simultaneously.

We have recently developed a new numerical simulation for light-matter interactions solving the Maxwell equations for macroscopic electromagnetic fields and the TDKS equation for microscopic electron dynamics simultaneously. We here outline this new scheme briefly. Further details of this approach are given by Yabana *et al.*[15]

6.1. *Maxwell + TDDFT multiscale simulation*

In describing dynamics of electromagnetic fields of light and electron dynamics simultaneously, we immediately face to the fact that there are two distinct spatial scales, the wavelength of the laser pulse which is typically a few hundreds of nanometer and the motion of electrons in a crystalline solid induced by the laser pulse which is less than a nanometer. To circumvent this multiscale nature of the problem, we introduce a distinction of macroscopic and microscopic scales as in the ordinary electromagnetism of macroscopic media and assume locality on the macroscopic scale.

In ordinary electromagnetism, locality in macroscopic scale is introduced by the constitutive equations. Namely, the macroscopic polarization field is a local functional of the macroscopic electric field. We assume that the same locality in space holds under strong fields,

$$P_\alpha(\mathbf{r}, t) = P_\alpha[E_\beta(\mathbf{r}, t')]$$

$$= \sum_\beta \int dt' \chi^{(1)}_{\alpha\beta}(t - t') E_\beta(\mathbf{r}, t')$$

$$+ \sum_{\beta\gamma} \int dt' \int dt'' \chi^{(2)}_{\alpha\beta\gamma}(t - t', t - t'') E_\beta(\mathbf{r}, t') E_\gamma(\mathbf{r}, t'') + \cdots .$$

$$(6.1)$$

We use this locality assumption in macroscopic scale to build up our multiscale theory for the coupled dynamics of electrons and electromagnetic fields. We introduce two kinds of coordinates, a macroscopic coordinate Z (one-dimensional, see below) and microscopic coordinates $\mathbf{r}_Z$ around each macroscopic position Z. At each macroscopic position Z, we calculate the dynamics of the Kohn–Sham orbitals, $\psi_{i,Z}(\mathbf{r}_Z, t)$. As discussed below, the locality is realized by assuming that the Kohn–Sham orbitals at different macroscopic positions, $\psi_{i,Z}(\mathbf{r}_Z, t)$, evolve independently.

We consider an irradiation of a linearly polarized laser pulse normal to the surface of a bulk solid, taking a coordinate system with the surface located in the xy plane at $Z = 0$. The polarization direction of the incident laser pulse is set parallel to x axis. This is a one-dimensional problem on the macroscopic scale, spatially uniform in the x and y directions. The macroscopic field is described by the vector potential, $A_Z(t)$, where Z is the macroscopic coordinate. The vector potential follows the wave equation as usual,

$$\frac{1}{c^2} \frac{\partial^2 A_Z(t)}{\partial t^2} - \frac{\partial^2 A_Z(t)}{\partial Z^2} = -\frac{4\pi e}{c} J_Z(t) \qquad (6.2)$$

where $J_Z(t)$ is the macroscopic current at point Z.

We next consider time evolution of the Kohn–Sham orbitals, $\psi_{i,Z}(\mathbf{r}, t)$ (we drop index Z of $\mathbf{r}_Z$). Since the macroscopic field changes little in the microscopic spatial scale, we may assume that electrons respond to a spatially uniform electric field in the microscopic scale. We thus use the same TDKS solver as that used in previous sections,

$$
i\hbar\frac{\partial}{\partial t}\psi_{i,Z}(r, t) = \left\{ \frac{1}{2m}\left[-i\hbar\nabla_{\mathbf{r}} + \frac{e}{c}xA_Z(t)\right]^2 \right.
$$

$$
\left. - e\phi_Z(r, t) + \frac{\delta E_{xc}}{\delta n}\right\}\psi_{i,Z}(r, t), \tag{6.3}
$$

where the macroscopic coordinate Z is regarded as a parameter independent of the local coordinate $\mathbf{r}$. From the KS orbital, one may calculate the microscopic current $\mathbf{j}_Z(\mathbf{r}, t)$ as in Eq. (2.10). Taking spatial average of $\mathbf{j}_Z(\mathbf{r}, t)$ over a unit cell, we may calculate the macroscopic current $J_Z(t)$ at each macroscopic position Z. We note that the current $J_Z(t)$ is entirely determined from the vector potential $A_Z(t)$ at the same position Z. This fact guarantees the locality of the response in macroscopic scale.

We now have the basic equations for coupled dynamics, Eq. (6.2) for the macroscopic vector potential and Eq. (6.3) for the microscopic electron dynamics. These equations may be solved as an initial value problem, the Kohn–Sham orbitals $\psi_{i,Z}(\mathbf{r}, t)$ at every position Z are prepared in the ground state and the electric field of the incident laser pulse is prepared in the $Z < 0$ region.

For the present formalism to be useful in practice, it should include the ordinary electromagnetism of macroscopic medium in the weak field limit. We may confirm it if we note the TDKS equation (6.3) describes dielectric response in the weak field limit as discussed in Eq. (3.1),

$$
J_Z(t) = \int^t dt'\sigma(t - t')\frac{dA_Z(t')}{dt}, \tag{6.4}
$$

where $\sigma(t)$ is the electric conductivity function in TDDFT. Since this linear relation provides the macroscopic current $J_Z(t)$ for any vector potential $A_Z(t)$, we may replace the TDKS equation (6.3) with the constitutive equation (6.4). Then Eq. (6.2) combined with Eq. (6.4) is the ordinary description of the macroscopic electromagnetic field in Maxwell theory.

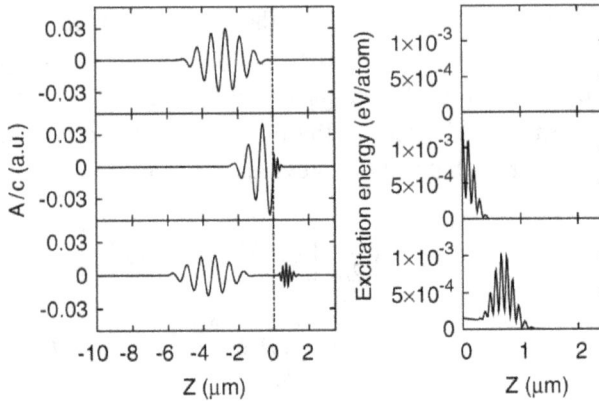

Fig. 11.   Snapshots of the vector potential divided by light speed, $A/c < 0$ (left panel), and of electronic excitation energy per atom (right panels) at different times are shown as a function of macroscopic position. The vacuum is at $Z < 0$ and the Si crystal is at $Z > 0$. Top panels: initial starting field with a pulse on left moving toward the Si surface. Middle panels: at the point where the middle of the pulse reaches the surface. Lower panels: the reflected and transmitted pulses are well separated. The maximum intensity of the incident laser pulse is set at 1×10^{11} W/cm^2. Taken from Ref. 15.

6.2. *Example: Laser pulse irradiation on Si surface*

As an example, we show a calculation of the laser pulse irradiating on a Si surface.[15] In Fig. 11, we show snapshots of the time evolution. In solving Eqs. (6.2) and (6.3), the macroscopic coordinate Z is discretized into 256 grid points with the grid spacing of 250 au for the medium region, $Z > 0$. At each grid point, microscopic TDKS equation (6.3) is solved. Computational aspects are mostly the same as those described in Sec. 2.4. The number of k-points grid is much smaller, 8^3 in the present calculation.

The initial laser pulse at $t = 0$ has a peak intensity of 1×10^{11} W/cm^2, the pulse length is 18 fs, and the average frequency is 1.55 eV. This is shown in the upper panel of the figure. The middle panel shows the field when the center of the pulse has just reached the surface. In the bottle panel, one can see a transmitted and the reflected waves. In the right panel, electronic excitation energy per atom is shown. The excitation is seen in the spatial region where transmitted wave exists. Although the laser frequency is far below the direct band-gap energy, the excitation is seen after the transmitted pulse went away in the right-bottom panel. This is caused by multiphoton excitations.

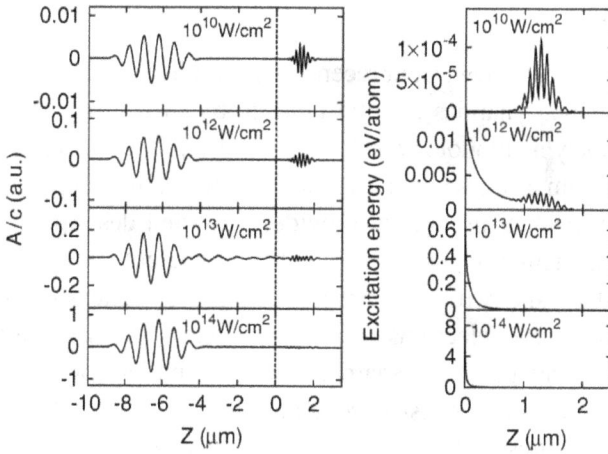

Fig. 12. State of the system after the incident pulse breaks into reflected and transmitted pulses for several different intensities of the incident laser pulse. Left panel: the vector potential divided by light speed. Right panel: excitation energy per atom in the Si crystal. Taken from Ref. 15.

In Fig. 12, we show reflected and transmitted electromagnetic fields at different intensity levels. In the left panels, the vector potentials are shown at a time when the transmitted and reflected waves are well separated. In the right panels, the electronic excitation energies per atom are shown in the Si crystal region. At the lowest intensity, the propagation of electromagnetic fields is well described by the dielectric response. Essentially all of the energy remains associated with the propagating transmitted pulse. As the incident intensity increases, the transmitted wave becomes weaker than that expected from the linear response. We also find that the central part of the transmitted pulse is suppressed strongly, producing a flat envelope of the pulse. In contrast, the envelope of the reflected wave does not change much in shape, even at the highest intensity. From right panels, above 1×10^{12} W/cm^2, one sees that most of the energy is deposited in the medium with just a small fraction remaining in the transmitted electromagnetic pulse. The deposition rate falls off with depth, as is expected from the weakening of the pulse. At higher intensities, the absorption rate greatly increases. At 1×10^{13} W/cm^2 and higher, the transmitted pulse is almost completely absorbed in the first tenths of a micrometer.

7. Summary

In this review, we reviewed our recent progress on a first-principles description of electron dynamics in solid induced by strong and ultrashort laser pulses. Our key methodology is the real-time approach for the evolution of the Kohn–Sham orbitals under a spatially uniform, time-dependent electric field. We show that the theory provides a unified description of optical phenomena, including dielectric function, coherent phonon generation, and optical breakdown. We further discussed an extension to describe coupled dynamics of electrons and electromagnetic fields, solving Maxwell and time-dependent Kohn–Sham equations simultaneously. It provides a general and comprehensive scheme for the light-matter interaction, although it requires quite large computer resources.

There are a number of issues which should be further extended, examined, and refined. The calculations presented here all employ an ALDA functional. It is well known that this simple choice has several deficiencies including a systematic underestimation of the band-gap. It is surely important to incorporate new functionals such as hybrid functional to increase quantitative accuracy, though calculations including nonlocal exchange require much more computational resources. The other aspects requiring further investigation include the treatment of collision effects. Although the TDDFT may in principle be able to take into account the correlation effects, it is not so obvious how to include collisional effects in the TDKS scheme. In describing coherent phonon generation, we assumed a classical dynamics for atomic motions while treating electron dynamics in quantum mechanics. The validity of this scheme, often called the Ehrenfest dynamics, may require a careful examination in general. While we consider it reasonable to describe atomic motions in the coherent phonon generation classically, there are many situations where quantum nature of atomic motions becomes important.

Still, the present computational scheme is applicable to a number of experimental measurements that make up of strong and ultrashort laser pulses. We expect that the real-time method will contribute to the development of the field through detailed comparisons with experimental findings. At the same time, quantitative comparison with measurements will eventually establish accuracy and reliability of this new approach.

References

1. J. L. Krause, K. J. Schafer and K. C. Kulander, *Phys. Rev. A* **45**, 4998 (1992).
2. S. Chelkowski, T. Zuo and A. D. Bandrauk, *Phys. Rev. A* **46**, R5342 (1992).
3. I. Kawata, H. Kono and Y. Fujimura, *J. Chem. Phys.* **110**, 11152 (1999).
4. M. Petersilka and E. K. U. Gross, *Laser Phys.* **9**, 105 (1999).
5. F. Calvayrac, P.-G. Reinhard, E. Suraud and C. A. Ullrich, *Phys. Rep.* **337**, 493 (2000).
6. X.-M. Tong and S.-I. Chu, *Phys. Rev. A* **64**, 013417 (2001).
7. K. Nobusada and K. Yabana, *Phys. Rev. A* **70**, 043411 (2004).
8. A. Castro, M. A. L. Marques, J. A. Alonso, G. F. Bertsch and A. Rubio, *Eur. Phys. J. D* **28**, 211 (2004).
9. Y. Miyamoto, H. Zhang and D. Tomanek, *Phys. Rev. Lett.* **104**, 208302 (2010).
10. C. A. Ullrich, *Time-Dependent Density-Functional Theory: Concepts and Applications* (Oxford University Press, 2012).
11. M. A. L. Marques, C. A. Ullrich, F. Nogueira, A. Rubio, K. Burke and E. K. U. Gross (eds.), *Time-Dependent Density Functional Theory*, Lecture Notes in Physics, Vol. 706 (Springer, Berlin, 2006).
12. T. Otobe, M. Yamagiwa, J.-I. Iwata, K. Yabana, T. Nakatsukasa and G. F. Bertsch, *Phys. Rev. B* **77**, 165104 (2008).
13. T. Otobe, K. Yabana, J.-I. Iwata, *J Phys. Condens. Matter* **21**, 064224 (2009).
14. Y. Shinohara, K. Yabana, Y. Kawashita, J.-I. Iwata, T. Otobe and G. F. Bertsch, *Phys. Rev. B* **82**, 155110 (2010).
15. K. Yabana, T. Sugiyama, Y. Shinohara, T. Otobe and G. F. Bertsch, *Phys. Rev. B* **85**, 045134 (2012).
16. Y. Shinohara, S. A. Sato, J.-I. Iwata, T. Otobe and G. F. Bertsch, *J. Chem. Phys.* **137**, 22A527 (2012).
17. G. F. Bertsch, J.-I. Iwata, A. Rubio and K. Yabana, *Phys. Rev. B* **62**, 7998 (2000).
18. Y. Shinohara, Y. Kawashita, J.-I. Iwata, K. Yabana, T. Otobe and G. F. Bertsch, *J. Phys. Consens. Matter* **22**, 384212 (2010).
19. J. R. Chelikowsky, N. Troullier, K. Wu and Y. Saad, *Phys. Rev. B* **50**, 11355 (1994).
20. N. Troullier and J. L. Martins, *Phys. Rev. B* **43**, 1993 (1991).
21. L. Kleinman and D. M. Bylander, *Phys. Rev. Lett.* **48**, 1425 (1982).
22. K. Yabana and G. F. Bertsch, *Phys. Rev. B* **54**, 4484 (1996).
23. M. E. Casida, C. Jamorski, K. C. Casida and D. R. Salahub, *J. Chem. Phys.* **108**, 4439 (1998).
24. A. Zangwill and P. Soven, *Phys. Rev. A* **21**, 1561 (1980).
25. T. Nakatsukasa and K. Yabana, *J. Chem. Phys.* **114**, 2550 (2001).
26. K. Yabana and G. F. Bertsch. *Int. J. Quantum Chem.* **75**, 55 (1999).
27. K. Yabana, T. Nakatsukasa, J.-I. Iwata and G. F. Bertsch, *Phys. Status Solidi (b)* **243**, 1121 (2006).
28. J. Paier, M. Marsman and G. Kresse, *Phys. Rev. B* **78**, 121201 (2008).
29. G. Onida, L. Reining and A. Rubio, *Rev. Mod. Phys.* **74**, 601 (2002).
30. C. Attaccalite, M. Gruning and A. Marini, *Phys. Rev. B* **84**, 245110 (2011).
31. T. Dekorsy, G. C. Cho and H. Kurz, *Top. Appl. Phys.* **76**, 169 (2000).
32. R. Merlin, *Solid State Commun.* **102**, 207 (1997).

33. Y.-X. Yan, Jr., E. B. Gamble and K. A. Nelson, *J. Chem. Phys.* **83**, 5391 (1985).
34. T. E. Stevens, J. Kuhl and R. Merlin, *Phys. Rev. B* **65**, 144304 (2002).
35. H. J. Zeiger, J. Vidal, T. K. Cheng, E. P. Ippen, G. Dresselhaus and M. S. Dresselhaus, *Phys. Rev. B* **45**, 768 (1992).
36. M. Hase, M. Kitajima, A. Constantinescu and H. Petek, *Nature* **426**, 51 (2003).
37. D. M. Riffe and A. J. Sabbah, *Phys. Rev. B* **76**, 085207 (2007).
38. K. Kato, A. Ishizawa, K. Oguri, K. Tateno, T. Tawara, H. Gotoh, M. Kitajima and H. Nakano, *Jpn. J. Appl. Phys.* **48**, 100205 (2009).
39. K. Ishioka, M. Kitajima and O. Misochiko, *J. Appl. Phys.* **103**, 123505 (2008).
40. S. S. Mao, F. Quere, S. Guizard, X. Mao, R. E. Russo, G. Petite and P. Martin, *Appl. Phys. A* **79**, 1695 (2004).
41. A. Q. Wu, I. H. Chowdhury and X. Xu, *Phys. Rev. B* **72**, 085128 (2005).
42. D. H. Reitze, H. Ahn and M. C. Downer, *Phys. Rev. B* **45**, 2677 (1992).

www.ingramcontent.com/pod-product-compliance
Lightning Source LLC
Chambersburg PA
CBHW050553190326
41458CB00007B/2019